THE POLITICS OF MATHEMATICS EDUCATION

STIEG MELLIN-OLSEN
Bergen College of Education, Norway

THE POLITICS OF MATHEMATICS EDUCATION

D. REIDEL PUBLISHING COMPANY

A MEMBER OF THE KLUWER ACADEMIC PUBLISHERS GROUP

DORDRECHT / BOSTON / LANCASTER / TOKYO

Library of Congress Cataloging-in-Publication Data

CIP

Mellin-Olsen, Stieg, 1939–
The politics of mathematics education.

(Mathematics education library)
Bibliography: p.
Includes index.
1. Mathematics—Study and teaching. I. Title.
II. Series.
QA11.M483 1987 510′.7 87–4485
ISBN 90–277–2350–8

Published by D. Reidel Publishing Company,
P.O. Box 17, 3300 AA Dordrecht, Holland.

Sold and distributed in the U.S.A. and Canada
by Kluwer Academic Publishers,
101 Philip Drive, Norwell, MA 02061, U.S.A.

In all other countries, sold and distributed
by Kluwer Academic Publishers Group,
P.O. Box 322, 3300 AH Dordrecht, Holland.

Printed in The Netherlands

To Nora

TABLE OF CONTENTS

EDITORIAL PREFACE

The development of knowledge is never easy. One doesn't want to go over old ground again, but yet one needs to establish the new in the context of the old. One is also anxious about the novelty of the ideas — are they new enough, or are they too 'way out' to be acceptable? In some fields perhaps these criteria are less important than in others. In education, I sense that 'novelty' is a tricky criterion, varying in value from society to society. In some societies the new ideas have to justify their adoption in the face to the old, tried and tested ideas. (Better the devil you know than the devil you don't!) In other societies the old ways have to justify their continuation in the face of the new, promising and exciting ideas. (I can't find a good proverb for this! Perhaps proverbs are all about preserving the past?) In any case, some people will argue, there is nothing new to be said about education anyway — the problems are the same and it is only the context which changes.

Mellin-Olsen develops the reader's knowledge through this book in ways that are both novel and challenging. Their novelty is not in question, judging by reactions to them which vary from "they have nothing to do with mathematics education" to "they concern everything that is done in mathematics education". Any ideas which produce these reactions are, clearly, challenging as well, and it is no accident that the book represents its own message.

Of course it's only a book, and like some of the young people Mellin-Olsen refers to, the reader may feel that it is "words, words, words" and so what? Well, the spirit of this book demands (and provokes) not only 'reading', but more importantly 'interaction'. Mellin-Olsen challenges your assumptions, confronts you with new perspectives and generally sets out to make you uncomfortable with your own knowledge. You become not just a reader but an inter-actor.

The book reflects the man. Those who know Mellin-Olsen will know his interactive style and should be delighted to find it in print. Those who don't know him will soon do so when they begin reading. He is a very honest and brave writer. Not the least part of his bravery is to put his ideas into English. Mellin-Olsen refers in his Foreword to my contributions to this project, but I should also say what a pleasant task it has been to interact with him about this work. As we often find in this international field of mathematics education, words and ideas about mathematics education don't translate easily from one language to another, and that difficulty is a powerful source of growth and development. I enjoy having my ideas challenged and my knowledge criticised. If you do too, you will enjoy interacting with this book. If you don't like having your ideas challenged you might not enjoy this experience, but I guarantee that you will never think about mathematics education in the same way again.

ALAN J. BISHOP.

FOREWORD

In September 1985 we heard about young people in Brixton and Frankfurt fighting with police in the streets. On the television news we regularly saw battles between young people and an armed police force. From other countries, such as South Africa, we heard daily about other fights due to simple civil rights being refused. For many, the world in 1985 was not a peaceful place.

In mathematics education we have arrived at a stage where we discuss personal knowledge, shared knowledge, the need for pupils to develop their own mathematical ideas and tools, in order to gain some insight into the power of mathematics.

It is difficult to see how personalised knowledge can be discussed without using notions such as politicisation, conflict and oppression, when the potential learner may be in the midst of bitter struggles for civil rights. At the same time it is not exaggerated to say that mathematics educators have not been at the forefront when it comes to politicising education.

In our seminar on Culture and Mathematics at the College of Education in Bergen we did not manage to do more than start a discussion on the theme, or look for consensus and contradictions on the issue. Paulus Gerdes, from the People's Republic of Mozambique, stated at the closing session that he was astonished how we, the European members of that seminar, were ignorant of the discussions and research carried out in developing countries. I agree with Paulus. I believe that the great leaps forward in mathematics education, or the didactics of mathematics, will be taken in the developing countries. One of the reasons I see it like that lies in the immense energy and potential for clear thinking which are released through successful liberation-struggles after centuries under imperialism and colonialism.

In what follows I attempt to build a general theory of the politicisation of mathematics education. Perhaps I should write "towards a . . .". This book is the result of a twenty-year long search to find out why so many intelligent pupils do not learn mathematics whereas, at the same time, it is easy to discover mathematics in their out-of-school activities.

It is by going into this issue in depth that I find general conceptions such as hegemony, oppression and resistance indispensable. I am back to Third World education and liberation, where the impact of such conceptions is so clearly seen.

It is trivial to say that a book with a title of the type "The Politics of . . ." cannot achieve an easy birth. The gratitude felt towards those who have helped is great. This is the first time I have been given the opportunity to write in detail about our work in mathematics education for a non-Scandinavian readership. I was welcomed by Dr. Alan Bishop at the University of Cambridge to spend two terms there, and thus became familiar with an international academic setting. The inspiration of this milieu, and the continuous encouragement and support from him have been indispensable for the completion of this book. His cooperation on my project included social welfare as well as a series of discussions about the manuscript. In many cases these discussions led to new considerations and an extended and deeper analysis of the topics I was working on. Not least, I learned a lot from Alan about the impact of culture on mathematics education, and it is due to his insight and experience that *culture* has received the dominant place it has in this book.

Thus in many ways Alan Bishop's commitment to promoting non-traditional ways of approaching mathematics education has made this book possible. In Cambridge I also received support from Dr. Marilyn Nickson who inspired my progress. Being a female mathematics educator trained in the philosophy of knowledge, she had a sound basis for supporting a project on the politics of education, which proved most helpful for me. I am also indebted to Professor Bent Christiansen for critical comments on my exposition of Activity Theory.

A large number of my colleagues and students in Norway have stood beside me and made this work possible by collective innovative work. Their contributions are reflected in the various projects reported throughout the book. In particular Einar Jahr, Marit Johnsen Høines and Ragnar Solvang have helped me to sharpen my definitions and analysis. One of Marit's contributions to this book is the analysis of the uses of language developed in Chapter 2. A series of examples and case studies in that chapter also "belong" to her. I am also indebted to another colleague, Steinar Hauge, for the illustrations.

I should also like to express my gratitude to the librarians at the Educational Library of the University of Cambridge for their friendly support. Libraries and their staff are important for any scientific project, and it is alarming to see how the libraries are among the first to be hit in times of economic cuts.

I do not know of anyone else who has been granted a sabbatical year or similar official support to explore the politics of a school subject. Nevertheless the British Council and the Norwegian Union of Authors of Science offered ample financial support, which made this project possible. I hope they did the right thing.

Bergen, September 1985 STIEG MELLIN-OLSEN

INTRODUCTION

Setting the Scene

> In these remarks on the schools, I do not try to be generous or fair, but I have seen what I am talking about, and I hope I am rational.
>
> Paul Goodman

1. I can easily remember them, those friends of mine, from the backstreets of Bergen during the mid-fifties. The night was usually dark and damp, and we would sit in someone's cellar, trying to get some warmth out of the tiny cigarettes.

"Hey, Prof," they would address me: I was the only one around that had sights on the gymnasium.

"Hey, Prof, I don't understand how you can cope, staying at that mad-house. We had another fight with that pig of ours today. He thrashed the three of us just because we set fire to those litter baskets."

"You idiot. You know that the pig will always be standing around the corner, watching you. That is his real interest in life — seeing you do it so he can get the chance to thrash you."

"I am glad it will soon all be over. Today I heard that I could become a printer's apprentice. My brother works at *The Chronicle*, you know. In the transport depôt. He has arranged everything. As soon as I have finished at that madhouse. . . . Only 22 days to go now."

"Printing! That's real work! I would have liked to do that. They have got a new rotary press, haven't they?"

"Yes, straight from Germany. They spent two weeks putting it together. Now it runs like a madman. 50,000 copies an hour."

"50,000 copies. That's bluff. Sheer bluff."

"It's no bluff it isn't! Almost a 1000 copies a minute. Spitting them out like bullets from a machine gun."

"Printing — that would have been something! That is the future. Composing the pages, making them look decent. Next morning everyone reading what you have produced. Your equations wouldn't have helped you there, Prof."

2. I met them again ten years later. I met them as a teacher of the lowest streams in a comprehensive school in Bergen.

"Hey, Teach," they would say. "Are you giving us those fucking equations of yours again? OK, Teach, you're not that bad, I'll work on them for a while. I am in such a good mood today."

After one of the rarer lessons, some lad would perhaps come up and say:

"Hell, Teach! If it goes on like this I think I can learn something at this school. Fantastic teacher! Do you think you could do it again?"

But in the next lesson we would be back on the trail again. There would be new equations, and I could feel sweat dripping down my back as I demonstrated the changing signs. We negotiated and discussed social matters during the last half of the lesson.

"You should see our pigeon hut, teacher. We have built it high up the hill-side. Eight square meters. Big enough for parties. Oak covering on the floor. We collected the wood from the wood-shop some nights ago. You should come and see it Teach."

"What about the pigeons then? Do you have any room for them?"

"Don't be stupid Teach. My best one was second in the race from Denmark last Saturday.

"Think about that, Teach! From Denmark. Over Skagerak. Straight back to the hut, it came. Beat the bad weather and everything. 878 metres a minute, average speed! That's something, isn't it? Straight from Aalborg. But I have trained her properly. I almost never lose a bird."

"A tenth year they say at home. They want me to do it. I want to myself too. Enter the grammar school. I must do that to get a job. Do they have maths at the grammar school? Of course they have. I must pull myself together. I am just so stupid. I don't understand a thing about the problems you give me. Do you give extra lessons, Teach?"

3. I still meet them. They are still the same . . . almost. I do not meet them as a teacher any more. I am now an outsider to their school, which makes it easier for them to talk freely. They do not defend themselves so much.

"Do you know what that idiot told us? We might be brain-damaged! He had learnt all about it on some course. He says it is a typical symptom, that's just the words he uses, a typical symptom that we move about when he is doing his fractions. What is it he calls it? Being Chinese? He must be mad, that man."

I knew that the man was not mad. He had just been on a course in special education and learnt that restless children should be called "hyperactive" or even "hyperkinetic".

"Better sitting in that flat than going to school! We went dead yesterday, the three of us. We smoked pot together. It was King Kong who turned up with it.

"King Kong's mother was taken away some days ago. It was terror it was — she was dead drunk. The police drove straight up to their front door and took her away. Everyone was watching, even the adults stuck out their heads to have a look. No one knows what will happen to King Kong now. His father has been away for quite a while.

"Work? You can ask, man. The Careers man said we would have a poor chance. Said it straight out, he did. Terror it is, saying like that — Sorry lads — you have a poor chance in obtaining any job after leaving school!

"He told us that the employment office would provide us with courses where we would learn how to work. It would be almost the same as having a job. I didn't understand a single word he was saying. Bluff it must have been, all of it. I would rather go to Copenhagen after Easter and have a big feast. That's what counts. Making life as pleasant as possible. Don't you agree, Teach?"

4. Starting as a mathematics teacher during the mid-sixties was a golden affair. There were the curricula reforms; there was Piaget, and all the mathematics educators who went ahead, inspiring the rest of us to walk in their footsteps. There were enthusiastic efforts around to transform mathematics education at all levels, from the widespread use of the drill method, to teaching based on conceptual and structural learning. The desire to focus on mathematics as such, to foster the growth of mathematical ideas among the pupils, was almost unanimous among mathematics educators.

At the same time great strides were being made in the political field. The slogan was one of democratisation, making the rich fields of theoretical knowledge available to a larger proportion of the population. In Norway and Sweden the social democratic governments went ahead, introducing the nine-year comprehensive school, without streaming, based on the idea that equal rights to education meant equal access to the same curriculum. The economy prospered and made it easy for those of us with the right ideas and the right education at the right moment to obtain generous grants. In this way even new materials could be produced and new groups of teachers could be persuaded to follow the golden road which would lead their pupils to the land of wisdom.

We got schemes called

— Mathematics without tears
— Mathematics with a smile
— Hey-mathematics
— My mathematics
— Our mathematics
— Mathematics for the majority

just to mention a few. And we saw some smiles, quite a few tears, and a disturbing amount of ignorance.

5. So there would be moments of hesitation. The smooth surface of educational reforms was also disturbed by some rough weather outside the primary and secondary schools. It was the time of the critics of the Frankfurt School. It was the time of Habermas and Marcuse, not to forget Marx. It was the time for the revitalisation of Socrates and Hegel, who had something to say about dialectics which could be connected with education.

Just that is worth a chapter in itself, as the pedagogues who solely relied on the dialectics of Socrates and Hegel proved to be a dead end for teachers. They merely examined the educational situation in terms of dialectics as a ritual, where the subject-matter of the situation played a subordinate part.

The Socratic dialogues are an exercise in how to respond to a learner. They do, however, provide a demonstration of creating a particular *form* (the learner's search for insight through the questioning of his master) embodying the *content* of the dialogue. It will be a major conviction in what follows that any dialectics in an educational setting which includes the learner as an agent credited with the right to influence the learning process cannot be separated from the content matter.

But the Hegel and Socrates of the late sixties lent support to those who were striving to reject authoritarianism. For a short time we also experienced power relations turned upside down by groups of students (later to be known as the 1968-generation) who operated a dictatorship from the grammar school level and upwards. All this was quite disturbing for a mathematics teacher who had already graduated within the safe walls of the Institute of Mathematics, where the major discussions had related to whether a negation-free mathematics would prove to be of significance or not. The revolutionary events were at a distance, but still there was an apprehension that they must in some way be connected with the emerging resistance among pupils that could be experienced at the primary and secondary level. The rejection of authoritarianism was also present there, although it was not recognised as such. It was more a confusion and a worry to the teacher — when the pupils started to question their curriculum.

Quite a few pupils were slow to obtain insight into the various contributions of set theory to algebra and geometry. This was so even when the methods of teaching were as prescribed by the experts in the field: constructive, inductive, intuitive, schematic, structural — to list some of the most-used labels.

It was even worse than this. Several of the pupils expressed their lack of motivation for the kind of stuff with which they were confronted. Some of the girls preferred to knit pullovers (beautifully designed patterns, by the way), and some of the boys would rather have filled in the pools. If the weather was nice some would ask pleasantly but persistently if they could go for a walk in the nearby park.

6. In my first term as a teacher I was naive enough to go and see my Headmaster, who happened to be President of the town's Labour Party, and tell him about my defeats and emerging frustrations. I really got a kick.

"We have fought for years against the bourgeoisie in order to establish a democratic school. And now we have to cope with you spoilt, privileged academics who believe you are the only ones to possess knowledge. I tell you: it is the best thing that ever happened to you, that you have to face these kids. So I will not listen anymore to such complaints.

That is — I really know some of these bandits you are telling me about; I shall go and have a word with them, so they will know for the future. Such opportunities we provide them with!"

But the rejection of the kind of mathematics that followed the official curriculum and its related examination, sometimes become unbearable. And the opinions were stated clearly, not only by the pupils, but also by their parents:

"Why don't you do percentages? It is terrible to see young lads come to work these days who are not able to do percentages. I myself don't understand a single thing you are doing. I am just not able to help my child out."

Of course there was the odd encouraging moment. We could investigate loci in a much more general context than the traditional Euclidean one, and the lower streams would discover dozens of them.

We could examine multiplicative structures such as those derived from chain letters, the spread of a rumour, and so on. Thus we could introduce the notion of a power and of exponential functions, and the pupils could be engaged in the kind of mathematical activity that would satisfy most mathematics educators at that time. It was a period where didactic experimentation was encouraged by all sorts of authorities within the educational system, a period which provided the mathematics educator with opportunities he had never previously had, and has not seen since.

So we had our good moments. The pupils gained operational knowledge all right. They could relate various concepts to each other in the proper way, and gain credit for it on the various tests we gave them. They gave evidence of operational learning, constructive learning, relational learning, schematic learning, call it what you like.

But the growing rejection by many pupils could not be neglected, and gave rise to discussions among many educationists. The telling signs that it could not be only lack of intelligence that caused the learning problems were present all the time.

"OK Teach. We have to do this. I know I must have my exam in order to go to the grammar school. But what we have to do is silly nonsense."

7. The inspiring and heartening writings of teachers such as Holt, Kohl and Kozol, to mention a few, described experiences similar to our own.[1] I was especially drawn to the ever-present phenomenon described in such detail by John Holt and Jules Henry, in which I as a teacher focused on the content matter in a particular situation — one in which most of the pupils would focus on specific strategies in order to obtain the correct answer.

This was easily revealed in mathematics, as the pupils repeatedly came up with fixed rules and final solutions when the teacher asked for reasoning, explanations, logic and structures. The pupil's tendency to handle mathematics without any concern for what they were really doing was quite disturbing (Are we to multiply or divide here, Teach?), especially as several of us as teachers tried to employ all the criteria which could be set up for sound, conceptual learning at that time.

Together with Skemp (1976) I did some research on this phenomenon in England and Norway. We gave the pupils a set of mathematical explanations from which they had to choose. The criterion for selection was that they had to pick explanations which showed why a mathematical principle was true. Our point was that we wanted to see to what extent they chose explanations which demonstrated how the mathematical principles worked in application (instrumental explanations). We tested several curricula in England and Norway in this way. Although there was some variation in the results, it was so small that it could say hardly anything about qualitative differences in the pupils' learning.

After this Skemp and I parted. He, as a psychologist of mathematics education, went further by investigating psychological formulae which could promote relational understanding (Skemp 1979).

I was more restless. I had previously carried out some similar research, evaluating various Norwegian curricula. In this research the pupils themselves had to produce explanations of a set of given rules and examples. After they had done this, I asked them to tick those explanations of theirs which they felt happy with, and put a minus sign against those they were not satisfied with. It was thus possible, in my view, to see whether a pupil was biased towards an instrumental or a relational attitude to learning.

Having scored their answers and made some statistics, I went back to the class to tell them what it had all been about. To do so is not only a question of simple ethics on the part of the researcher, since the classes and their teachers volunteer for the research. The pupils will often respond to the test without even knowing what they are actually responding to. Going back to the pupils to discuss the test with them also provides the researcher with useful information about how the pupils perceived the test.

So it happened in this case. During the discussions with the pupils it turned out that they had often looked for quite different qualities in their explanations from those that were my concern. Sometimes, and I could verify this afterwards by going back to their written replies, they would rate an instrumental explanation positively if they had obtained some impressive formulations which looked like those frequently used by their teacher. At other times they would be unhappy with an explanation which pointed to relational understanding but was not communicated fluently. They would also be satisfied with an explanation which was presented in grand terms even if its content was pure nonsense (Mellin-Olsen, 1977, 1981).

8. Looking at these phenomena, supported by investigators of the classroom such as Holt and Henry, I had to consider the relational and instrumental responses as two aspects of the same reality: they were examples of how the pupils interpreted the context of their educational situation. I found a theoretical solution of this some time later when I met the use of "instrument" in the field of the *economy* of education. Altvater (1974) used the term in order to describe school as an objective tool for its pupils, with which they could obtain the necessary qualifications for labour in order to make a decent life as adults. School thus functioned as an *instrument* for its pupils.

Accepting this view, it was easy to see how it had to produce a rationale for learning which was not related to whether knowledge was structural or not, but rather was tied up with the various forms of appraisal pupils could get from their school.

The failure of cognitive psychology and most of the learning psychology of the seventies was that these sciences had no tradition of embedding the context-level of behaviour in their theories. Thus, the theories could not explain how behaviour could be interpreted in terms of types of learning situation. Mind you, I am now writing about the context to which the pupil relates the situation and how this affects behavioural patterns. I am not talking about the context which the experimenter or the educationist expects or hopes to be present and recognised by the pupil.

In such a matter, of course, psychology is not able to examine the dialectics between the interpretation of context and the resulting behaviour, a dialectics which is essential for the understanding of learning. How can the interpretation of context develop and change by varying the content of the learning situations, and conversely? We look for learning theories which regard the learner as an active, interpreting, evaluating, cooperating or rejecting participant in the learning situation, whose behaviour is significantly influenced by such processes.

9. For some reason, however, the ideology of those in charge of

remedial programs always seems to imply that it is the pupil, rather than the curriculum he has been exposed to, which has to be cured. Now in the eighties we experience the blossoming of behavioural modification theory. In the 1960—70s we saw several educational programs of *compensation* practised. The pupils got labels such as *functionally disabled.* It soon turned out that the largest sub-group consisted of those who were *socially* disabled, at least on the Norwegian scene. Schemes for teacher training were set up and a considerable proportion of the teachers attended in-service courses, thus achieving a post graduate degree in special education.

Special remedial programs were designed in order to cope with the situation. The dominating paradigm for these was one of intervention: when the pupils were not able to learn at school, something had to be wrong with their readiness, and the causes for this had to be found in their nearest social environment, that is, in their culture. Coleman (1968) is an example of such thinking. When summarising the discussions about the concept of equal educational opportunity, he starts off by criticising the four point ideology which was basic to the extensive postwar educational reforms in the Western culture:

1. The provision of free education up to a given level, which constitutes the principal point of entry to the labour force.
2. Providing a common curriculum for all children, regardless of background.
3. Children from diverse backgrounds attend the same school.
4. In the same school, the children will make use of the same teachers and the same facilities.

Coleman's conclusion is that the equality of education should not be measured by the equality of educational inputs, but rather by the intensity of the school's influence relative to the external divergent influences. So compensatory methods of education became for a while more predominant: a supposedly poor language had to be enriched, poor social contacts between parents and child had to be taken into account and so forth.

Stone (1981) describes the compensatory programs of that time as

> a form of "resocialization". The family and the social group is regarded as being (a) incapable or (b) unwilling to inculcate the collective social values into the young, and organized social institutions have to assume this task. Compensatory education as "resocialization" is directly related to this discussion because inevitably socialization is bound up with self-concept and self-image.
>
> Ibid., p. 83

In the US the Coleman Report (1966) gave sociological support for the compensatory theory, and in England the Newsom Report (1963) (focusing on the deficits of the pupils) and the Plowden Report (1967) (the deficits of the family) did the same. They built their investigations on the model that school had to be disregarded when failure to learn occurred, as such failure was usually tied up with lack of support from the home. It was the family which basically influenced children's learning. If those nearest the children, i.e. their parents, were not friendly towards and understanding of the school's efforts and goals, and did not support these, the children would necessarily fail to learn in school. Thus the resulting strategy for the school had to be to encourage the families to give the school a helping hand rather than neglecting the school's efforts.

10. If only it had been that simple. But as Baratz and Baratz (1970) point out, you cannot solve problems by correcting deficits when the deficits are not real.

In the US pedagogues argued about the success of the extensive Head Start Project. This project was the main thrust of President Johnson's War Against Poverty. It was directed towards *families* in order to help them to bring up their children in a relevant way, in order to prepare them for what awaited them at school.

As Volume 40 of the Harvard Educational Review richly demonstrates, the evaluation of the Head Start Project caused quite a provocative discussion to ensure, about what the project had achieved. Studying this discussion, one has again to wonder why hard-data researchers in the field of education are intrigued by differences in figures rather than their magnitude. Among the defenders of the evaluating procedure, we find Cicirelli et al. (1970). They report tables which show that one of the few significant differences at the 0.05 level is the one found for the Stanford Achievement Test, grade 1, where the Head Start Children obtained an average score 44.42, while the children in the control group obtained 41.13.

It is hard to see what contributions lengthy discussions on such matters can make to curriculum development. In Scandinavia Gjessing (1974), one of the leading special educators there, summarized the Scandinavian research which evaluated schemes of special education. He found that despite the fact that one could not deny that some children had benefited from special education, one could not neglect the evidence that several had suffered from it (as by stigmatisation). Furthermore, on average there was no evidence to show that children showed any specific development which could be related to the remedial programs in which they participated. This was the case whether it concerned IQ scores, skills in the major subjects or social adaptation. These results related to all IQ groups. The interventionists worked on the assumption of a social pathology of children's culture. But whatever definition one wishes to give the concept, a culture is so deeply rooted in history, ideology, religion, not forgetting the struggle to cope with daily needs, that it would be astonishing if its members would give it up overnight for the sake of the demands of school.

Consequently it seems worth attempting to construct theories of mathematics education and instruction (or of education in general) in which *culture* is a basic concept, rather than making new attempts to neutralise cultural influences on learning.

11. I especially remember Tommy, whom I worked with for almost a year. We did mathematics following a new scheme specially devised for special education by some teachers who had done "modern" mathematics.

Tommy took three medicines daily. One for his epilepsy (which I doubted as I learned to know him), one for his sinus problems, and one tranquilliser. He was diagnosed as a hopeless case as regards normal social life, being in the IQ range 60—70.

According to his teacher he was to do intersections. So Tommy was confronted with this set of numbers from the 2×-table, and the set of numbers from the 3×-table, and he was to pick out those he saw in both places.

Most of the time he grinned at me.

"Most idiotic thing I have done. See if I find 6 in that ring? I am thirsty."

He would go to the washroom as I nodded in agreement, trying to ignore the protesting face of his teacher.

Tommy had another lesson to teach me. We did sums. 2+1, 1+2, 2+3, 3+3.

There was counting on fingers and there was lots of yawning. The 3+3 caused severe trouble because Tommy did not know how to count his fingers when he went above 5.

But then all of a sudden: 3+4.

"That is seven", Tommy would say, returning to nailbiting and fingercounting when facing 4+4.

"5+4"?

"Nine" Tommy would say. "Have we to do many more now?"

"Wait a minute Tommy. How can you find these so easy when the others are so difficult?"

Tommy shrugged his shoulders, paused, and gave his answer.

"You know, Teach," he would say. "You know I have to play my role."

"Which role?"

Silly me, how could I expect an answer to that one.

"Where else do you play your role?"

"I don't know. Perhaps on the bus. Can I go now?"

I pursued the role hypothesis during the remainder of our time together never obtaining any firm answer to it. But Tage Werner tells almost the same story coming from his work at a similar school in Copenhagen. He and his pupil had done some laborious and tedious work in the classroom and went out to buy some cigarettes. The boy checked his change, seeing that everything was all right.

"Is it?" Tage asked.

"Of course, you can see it for yourself."

"But in there?" Tage nodded towards the school.

"Oh, that is in there."

This is not the place for speculations, generalisations or the introduction of role-theory. Rather is it a place for hesitation, confusion and some questioning.

12. The contradiction which followed from the conviction about the paradigm of looking at the pupil when examining the causes for educational success, or lack of such, could not be hidden any longer. What a relief it was when high-status researchers changed their research paradigms.

The Norwegian psychologist, Jan Smedslund, who made such supreme experiments with Piaget in Geneva, wrote the following:

> When I meet a little child I always take it for granted that, within its restricted range of activities and on its own premises, it is logical, and my problem thus becomes to understand what its behaviour means and thus to understand its existential situation.
>
> As long as the psychologists of the Piagetian tradition concentrate on logic as a variable factor (e.g. those who conserve and those who do not), and only to a little

extent pay attention to the problem of recognizing how the child interprets the situation and the instruction, I believe they are making an epistemological error and are at variance with daily life as well as all useful practice.

Smedslund 1977, p. (transl. S.M.-O.).

Another Norwegian, Hoem (1972), had his thesis on the learning of the Laps accepted as a doctorate in education. The Laps are an ethnic minority in the North and those living in Norway have to go to Norwegian schools in order to receive a Norwegian education. Hoem revealed that for several Laps it was a good thing that their children did poorly at school. It meant that there was a pretty good chance that their children would not socialise in Norwegian culture and would be "saved" for their own, heavily threatened culture. It was rather startling for any of us by then, as teachers, to meet an argument of the type that "for some it can be a bad thing to be successful at school."

Similar research followed in other parts of the world. It all signified the importance of the pupil's attitude to the learning situation, in particular, and to school in general, which in the end proves decisive for the resulting learning. There was of course the work of Labov (1972), who saw how black children of the New York ghettos maintained a defensive attitude when they faced the psychologist who was testing their vocabulary.

Labov's research points to another important aspect of the compensatory programs, which I have already mentioned briefly: it is not possible for the sociologists or pedagogues to judge cultures foreign to their own as being deprived, since they actually, for the time being, do not know what to look for. In other words, in order to play fair to the culture, one should at least examine it from the inside: not only to find out about vocabulary, grammar, knowledge use, but also the basic rationale governing learning behaviour.

This is, of course, a call for the approach of the social anthropologist. It requires the method of participating observation, it is time-consuming, and it will often be a demand on the researcher that he chooses sides, giving up some of the academic ideals of neutrality. This is exactly what Hoem did when carried out his research *among* the Laps, and it was what Labov discovered in connection with his research on "non-standard" English: the language he studied possessed both logic and richness.

It was what we found in Bergen when we talked with 45 worker families about their children's schooling (Mellin-Olsen and Rasmussen 1976). We were familiar with the kind of social contexts into which we intruded, we had no fixed list of ingenious questions, and we gave careful support to the statements the parents made when they questioned the daily routines and practices of school. Still, most of the families defended themselves in this situation and it was only about fifteen who displayed fully their inner opinions and experience. The evidence was harsh, and we called our book *School's Violence.*

I should like to draw attention to one particular point which supports the claim for the inclusion of *culture* in a theory of instruction. Bergen is a town which is the prototype of the Scandinavian social democratic dream. There are very few rich people, and few really poor people. Inequalities in income, housing, opportunities for holidays, sport, cultural activities etc., are relatively small compared with what one might find in other European towns. Still the cultural conflicts between working class families and those who represented the school were evident. Thus in many cases the working class families found it worth defending themselves from school, as they usually felt themselves to be socially inferior to those representing the school: the headmaster and the teachers. Such conflicts not only concerned the social aspects of schooling, but also had relevance to cognition and perception in fields which are basic to mathematics learning.

13. The works of Willis (1976, 1977) demonstrate that it is not necessarily recognised as a defeat for working class boys to get working class jobs. In Willis's sample from the Midlands in England, it is not sad when the "lads" fail at school. On the contrary, Willis found that working class boys actively rejected school, it was a choice which was an act of freedom and independence. Willis's great contribution here is that in a profound way, he describes the transition from school to work for a group of working class boys, showing how they were determined to stay working class, how this determination was affected by the school, and also how it affected their attitudes towards school.

But I find his picture too simple. The various messages that reach working class culture are so different and conflicting that the story of the pride of *being* working class cannot be generalised.

We can easily discover conflicting ideologies within the working class. The phenomenon of Saturday school in England is one example. West Indian families participate eagerly in such schools, not only to obtain basic skills which the ordinary school fails to provide, but also in order to improve their chances for educational careers within the context of the British educational system (Stone 1981).

Hoem's research among the Laps was related to one group among the Laps, those who wished to retrieve their culture. There are equally many who want to assimilate Norwegian culture and live a Norwegian life.

In its very depth mathematical knowledge is not independent of the cultural contexts within which it is developed. Nor is the use of this knowledge. Mathematics, like knowledge in general, is developed within societies in which people cope with daily needs, where someone is in a position to plan ahead and where others do not need to reflect on bread and butter at all.

A society can base its production for survival on primitive farming or on the use of high technology. What is considered to be *important* knowledge in a society is not independent of the way people live their lives; how they produce, spend their leisure time (if any), communicate and so forth. I am not going to locate any precise concept of *mathematics* as a field of knowledge biased by culture in what follows. I will rather examine which knowledge *material* in a particular culture is feasible for mathematisation in order to allow exploration in mathematics.

Mathematics does not start with such concepts as *variables, functions, mappings, groups*. Rather it starts with counting, measuring, comparisons and so forth.

We know that what is counted and how it is counted differs from culture to culture. It is the same with measuring, the uses of geometrical shapes, the use of relations and other activities underlying what has acquired the status of *mathematics* over the last 2000 years.

I shall document *folk mathematics* (Section 1.1.2.) as knowledge biased by culture (or social class). I shall furthermore discuss Pinxten's findings in the context of the anthropology of space (Chapter 3). It is a matter of definition whether *folk mathematics* is to be defined as mathematics or not. To what extent folk mathematics is recognised as important knowledge is a political question and thus a question about power.

My aim is thus to argue that the different uses of mathematics in various cultures can be decisive as to whether the members of one culture learn the mathematics of a curriculum or not. Such a position implies that, as educationists, we not only have to include the cultural aspects when developing a curriculum, we also have to realise the impact of possible conflicts between the various cultures to which the pupils relate. We mostly have to cope with this problem in the context of the non-segregated classroom, that is, in terms of comprehensive education.

This is how it should be, if school is to be a fair image of the world outside it. The task is difficult, and we still have little experience and almost no educational theory of how to approach solutions. To no small degree this is the case in the field of mathematics education.

15. My reader should now have some picture of the kinds of perspective I shall use when discussing and hopefully developing mathematics education. One more goal still has to be mentioned. It is the one which may be the main motivation for writing this book. The future for young people of most of the world in our time does not look too good. In the UNESCO report *Youth in the 1980s* we read

> The key words in the experience of young people in the coming decade are going to be: "scarcity", "unemployment", "anxiety", "defensiveness", "pragmatism" and even "subsistence" and "survival" itself. If the 1960s challenged certain categories of youths in certain parts of the world with a crisis of culture, ideas and institutions, the 1980s will confront a new generation with a concrete, structural crisis of chronic economic uncertainty and even deprivation.
>
> UNESCO 1981, p. 17.

Our feeling is that there is increasing neglect of children by their nearest and dearest and that the school has take care of several aspects which previously were dealt with by the family. Thus social education plays an important part in the total activity of the school in a number of ways. But as it is usually practised, social education is independent of the education in the various school subjects. Social education is usually about housing, clothing, nutrition and social contact; it is not concerned with the teaching of language, mathematics or the social sciences.

When crisis hits society, there will be marginal groups which are more vulnerable than others. Pensioners and handicapped people are examples. Youth is another. The UNESCO report (1981) stresses how even groups among the youth are especially vulnerable:

> Young women, both in the industrialized and developing countries, will certainly feel the impact of the crisis as a blow to recently achieved but still fragile recognitions of

equal rights. Competing for scarce employment opportunities with their male counterparts, women are almost unavoidably disadvantaged. This will hold true for women at almost every social and educational level.

. . .

Ethnic minorities among the youth, and of course, the physically or mentally handicapped are also especially subject to the loss of employment and educational status inherent in a situation of austerity and unemployment.

Ibid., p. 20.

Rather than compensating for a difficult life-situation by drug use or hooliganism, we would like to see young people being constructive, inventive and forceful.

Children and young people should thus be in the possession of the means to communicate and give evidence about important features of their lives. They should be able to document and contribute to the solution of their problems, involving their fears, hopes, needs and demands. Language education is probably the most important subject in this respect. It is my conviction, however, that mathematics education can be as important. Mathematics is also a structure of thinking-tools appropriate for understanding, building or changing a society. We have few traditions to draw on. But somehow we have to make a start in order to explore the possibilities of our subject.

CHAPTER 1

ACTIVITY THEORY

> When I was a young man, I went to the people, to the workers, the peasants, motivated really, by my Christian faith . . . I talked with the people — the pronunciation, the words, the concepts.
>
> When I arrived with the people — the misery, the concreteness, you know. But also the beauty of the people, the openness, the ability to love which people have, the friendship . . .
>
> Paulo Freire

1.1. FROM PIAGET TO VYGOTSKY

In this first section I am going to prepare the ground for an exploration of Activity theory. The theme of this section is the paradigm that the learner *always* has some important knowledge which is significant for the learning process, knowledge which should thus be recognised by the curriculum maker. This paradigm will be central in the exploration of the theory.

1.1.1. *On Piaget*

Very few, perhaps no, psychologists have had such an influence on education as Jean Piaget. His grand theory about the development of intelligence in the child and the adolescent and his analysis of different kinds of knowledge and their acquisition were most welcome at a time when pedagogues were in need of theories which could bring new insight into problems concerning learning and teaching.

It was mainly Piaget's paradigm that the learner constructed new knowledge on the basis of his activities which was of interest, as pedagogues were very worried about the prevailing use of drill methods in (mathematics) education. Piaget's epistemological base is that knowledge, including intelligence as general knowledge, exists in the context of actions, such as manual actions and mental actions. Knowledge is thus constructed as the result of the individual's actions on his world.

Piaget's theory is thus a theory of actions, or an activity theory: development and learning arise from the activities of the individual, and how he or she constructs knowledge on the basis of these activities. Today we can see that pedagogues like those who worked in the field of mathematics education, underrated the depth of Piaget's analysis of *activity*.

It was the neglect of the content aspect which failed. Piaget himself would say that the neglect occurred when the analysis of a syllabus restricted itself to logical and mathematical aspects of knowledge, disregarding the psychological side.

Psychologists and mathematicians such as Dienes (1964, 1967) and Skemp (1971) built their theories of learning mathematics more or less disregarding content and its relation to the learner. A series of mathematical textbooks and 'Piaget For Teachers' books emerged, based on activities involving one-to-one correspondence and other conservation tasks, without any concern for *what* the object of the activities was and what the pupils thoughts about them could be.

Piaget would reject such an approach. In quite a modest book, rarely quoted by educationists (I have found none who refer to it), he stresses the difference between the psychological concept of number and the mathematical. In the following quotation he reveals the weakness of solely exploiting a mathematical concept of number:

> Russell and Whitehead's famous example of equivalence classes makes a correspondence between the months of the year, Napoleon's marshals, the twelve apostles, and the signs of the Zodiac.
>
> In this example there are no qualities of the individual members that lead to a specific correspondence between one element of one class and one element of another.
>
> . . .
>
> When we say that these four groups correspond to one another, we are using one-to-one correspondence in the sense that any element can be made to correspond to any other element. Each element counts as one, and its particular qualities have no importance.
>
> Each element becomes simply a unity, an arithmetic unity.
>
> Piaget 1970, pp. 36—37

During the 1970s Piaget's paradigm about the necessity for activity was extended. Mathematics pupils should direct their own learning activities to a certain extent. In other words, children who were permitted to pose their own problems and to construct their own algorithms, not necessarily the standard ones, were considered to be in a favourable learning situation. This idea brought a new dimension into the relationship between the pupil and her (or his) curriculum. It was no longer sufficient to provide the pupil with concrete experiences from which she could construct her mathematical concepts, as in the tradition developed by Dienes, Skemp and their followers. The new claims were for learning situations which offered the pupils opportunities to move on their own, posing their own problems and questions.

The works of the American educator Robert B. Davis are examples of such an approach.[1] So are the guidelines and practices developed by IOWO in Holland and many other places in Europe. Christiansen (1983) calls it *Self-activity*. I shall discuss an IOWO project in detail in §1.3.4.

There are still some relationships missing from didactical theory based on Piaget's general epistemology. I miss the pupil's evaluation of the kind of learning situation she is confronted with. I miss recognition of the fact that she is in a position to reject the kind of activity she is invited to participate in. I miss the relationship between her own judgement of the educational situation she is part of, how this affects her learning behaviour, and how all this relates to the design of the educational situation.

Such a relationship is a dialectic between two levels: the judgement of the particular learning task she is confronted with and her meta-knowledge about the subject, telling her that the tasks she is usually confronted with prove to be valuable to her (or not). These two levels influence each other dialectically. I pursue this dialectic in §1.2.2 and §4.3.

I miss the conception of *communication* or *dialogue* which is object-oriented and in which both teacher and learners participate. This inclusion of communication as a part of Activity will be made in §1.2.5.

We have seen a series of inventive and ingenious projects which accord with the principle of self-activity. Theoretically, however, the pupil has usually been considered as one who reflects on the mathematical content of the situation, and not *about* the learning situation. It is thus the theoretical consideration of the significance of the various context-levels of the instructional situation that I miss.

I shall thus make an attempt to consider theoretically the relationships between the pupil and the learning situation in terms of the pupil's evaluation of that situation.

1.1.2. *Folk Mathematics*

Piaget's original descriptions of the formal operational stage were too general to be used as a tool for building a theory of instruction. It was easy to interpret "slow" learners in mathematics, age group 12—14, as not having developed to this stage. When a pupil does not understand the principles of the proof demonstrating that a quadrilateral in which pairs of opposite sides are equal is a parallelogram, he probably does not have the ability to reflect on the basis of an assumption (not being able to hypothesise that the quadrangle which everyone could see was a parallelogram was not).

But the same young people who could never understood the point of such a proof succeeded in similar tasks in the context of manual work, as when they built the frame of a hut.

Piaget became aware of the generality of the formal operational stage. In another short paper which I have seen no-one else in the field of mathematics education refer to, he states that he is dubious about the formal operational stage (Piaget 1972). It seems that the ability to think ahead and use assumptions as a conducting means for reasoning, cannot be described generally. This kind of logic was more likely to relate to a particular profession, working experience, culture and so on. This statement is remarkable, as it is one of the few occasions on which a psychologist admits that one of his key concepts is actually culturally biased.

Lancy (1978) in the context of his research in the mathematics of Papua New Guinea argues similarly: the progress from concrete operational thinking to formal operational thinking is a progress of culture rather than of the individual.

One implication of all this for our work in Bergen was to study what Maier (1980) calls *folk mathematics*. Another term for the same notion is *colloquial mathematics* (Dörfler and McLone 1984). It is the way people outside the established society of mathematicians use the subject. Among children, we can observe folk mathematics in activities involving games, gambling, buying and selling. We also observe folk mathematics in use in building, construction and design, as I shall demonstrate below. The idea behind exploring such mathematics is the striking contrast between what the school rejectors could master of school mathematics and what they mastered of intelligent tasks outside school.

Linguists such as Bernstein (1975), Rosen (1972) and Labov (1972) support the view that success in learning can be dependent on factors studied in the sociology of knowledge and sociolinguistics. Their research demonstrated that lack of intelligence or knowledge was not the main problem of the underachievers, it was rather a problem of language *differences*. It became clear that it would be worthwhile to examine the discontinuity between the knowledge forms which the pupils possessed and those knowledge forms usually awaiting them at school.

The question became how to bridge the gap. We could, for instance, observe that boys and girls knew several ways of producing right angles in practical situations.

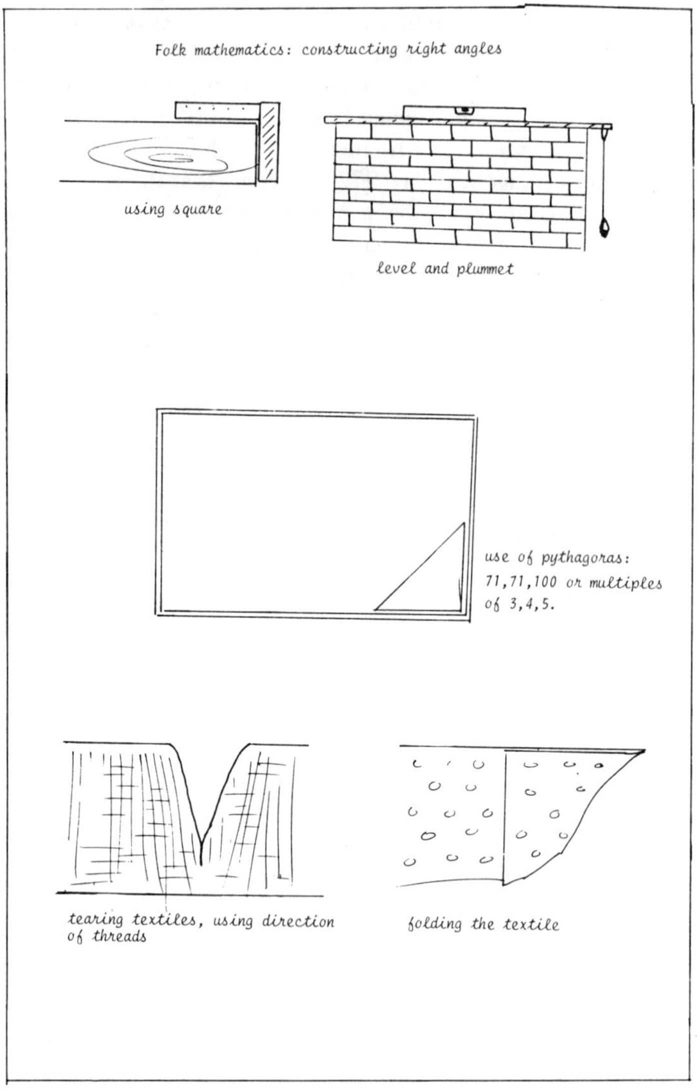

Fig. 1.1.1. Folk Mathematics

We could observe the same pupils produce results like this, using compasses and rulers in the context of the textbook problem: "Construct the perpendicular from point P to line m":

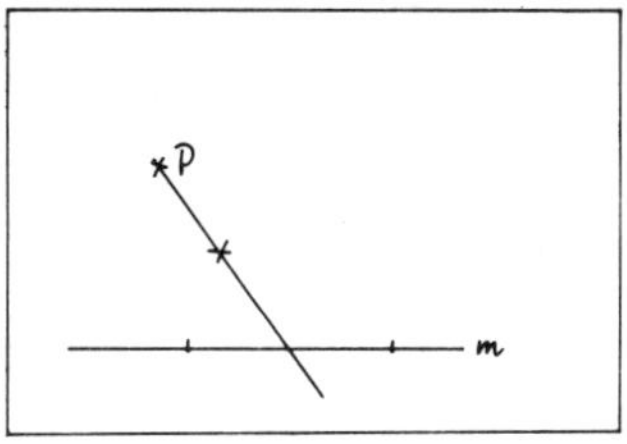

Fig. 1.1.2.

Similarly, the kids in Bergen, being used to building timber huts for their private activities, had a functional knowledge of how to saw wood:

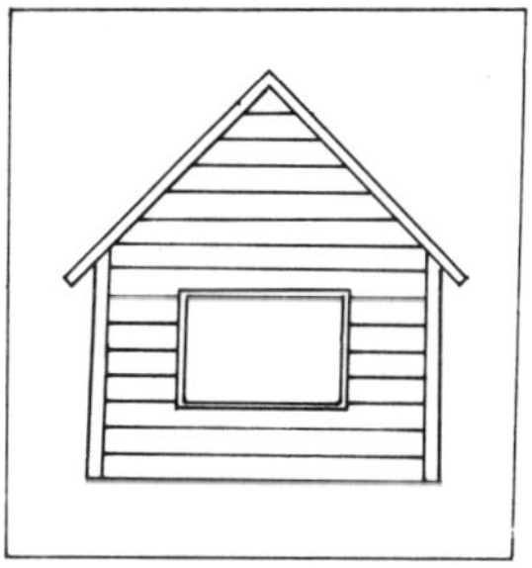

Fig. 1.1.3.

In school they would face the following problem:

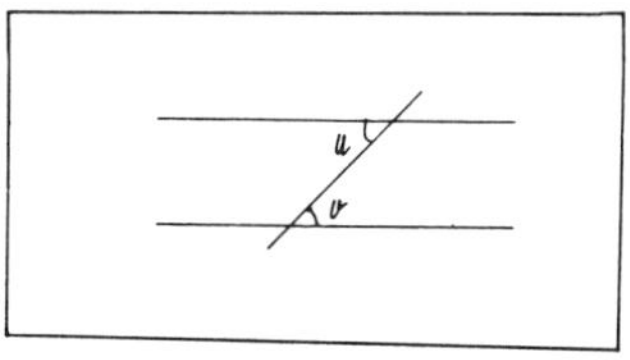

Fig. 1.1.4.

"Tell us why $\angle u = \angle v$".
The expected answer would be:
"Because they are corresponding angles between parallels."

The kids would have no idea about the word "corresponding". Furthermore they could not see angles as being *between* either, although carefully drawn with coloured chalk.

They came so close to Euclid, but they did not discover it; not did their teachers.

The important theoretical problem this raises is where is the location of a concept of *mathematics*. Does the boy or girl who can saw a piece of wood along a given angle necessarily possess mathematical knowledge about angles? Do we claim the ability to construct by means of compasses and ruler? Or do we regard a formalisation like

$$\angle a = 90^\circ$$

as a criterion of some mathematical knowledge?

Is it the shared, thus objectified knowledge as opposed to individual, subjective, knowledge which has the potential to acquire status as mathematical knowledge? Stress has already been laid on communication as a part of Activity, and thus knowledge. So mathematical knowledge will at least be shared knowledge, interpersonal knowledge.

I do not, however, see it as important for my purposes to contribute any strict definition of *mathematics* to exclude or include certain activities *as* mathematics. Nor shall I discuss the same. Whether the design of a skirt pattern involves a mathematical activity or not cannot be a crucial problem for a social theory of mathematics learning.

My argument is that such a design, as a piece of folk mathematics, is an activity which creates *material* for what most educators would accept as mathematics learning. Such material is *intellectual* in its form. Furthermore it builds on activities within a specific culture.[2]

Intellectual material will consist of experiences appropriate as a basis for further theoretical learning: abstractions, generalisations, relational thinking and so forth. The activities of folk mathematics point to the existence of such material.

Anthropological research which explores the qualitative and quantitative distribution of folk mathematics will thus provide the mathematics educator with material from which she (or he) can profit when she (or he) builds a curriculum. Whether such material passes the criteria set by a definition of mathematics does not, as far as I can see, have any important implications for a social theory of learning.

The material will be considered as an important prerequisite for the development of mathematical knowledge in its purest form. It thus coexists with formal mathematics. The importance of the concept of *material* thus arises from the argument that it is worth exploring it in order to take advantage of it when designing a curriculum for a particular group of pupils.

1.1.3. *Bruner's Modes of Representation*

The field of folk mathematics is extremely rich and rewarding for those who care to investigate it. It obviously calls on the methods of social anthropology. Some of our observations from Bergen will be reported in more detail in §3.3.5. In particular, the female students at the College of Education contributed in an important way. They had experienced frustration and suppression in mathematics at the grammar school, and had little faith in the subject or in their own capacity to learn it.

As soon as they realised that knowledge of their own also had value as material for the learning of mathematics, their attitudes towards the subject changed in a positive direction.[3] They drew on their experiences from sport and the household, especially soft handcrafts such as knitting, sewing and weaving, which hold a strong position in Norwegian female culture. Some feminists argued that such content for school mathematics represented sexism. It was the same kind of argument put forward by politicians who promoted an education which would free the worker from being working class and the farmer from staying a farmer, in order to gain a flexible society without class barriers.

The point missed by such an argument, however, is that in order to start the learners in question off towards liberating knowledge, the harbour of knowledge from which they depart has to be familiar to them. The thesis is thus about continuity: building generalising knowledge on the basis of the intellectual material present in the daily activities of the learners. In the long run, knowledge cannot be experienced as alien knowledge such that its only quality is its importance as school knowledge.

By introducing some mathematical knowledge to pupils by connecting it with their particular knowledge culture, it is possible for them to study the basic ideas it represents. Then it may be possible to study these ideas in a general way so that the liberating power of the knowledge may be recognised. A particular context in which knowledge is explored, such as one connected to folk mathematics, can thus be a prerequisite for liberation from the culture (if this is the ultimate goal), rather than a further attachment to it.

Bruner (1966) is helpful here. His stress on representation of knowledge by various modes makes sense. He describes three such modes, the *enactive, iconic* and *symbolic*. Bruner constructed his "modes of representation" for another purpose than mine. He is an educational psychologist who wanted to examine *development*, and his hypothesis was that these three represented such: first enactive, then iconic and finally symbolic.

What was interesting for my fellow teachers and me was that Bruner's conceptual construct fits in nicely with how our pupils in the comprehensive school demonstrated their knowledge. I related the enactive, iconic and symbolic modes of representation to knowledge possessed and represented by respectively (1) *manual work*, (2) *picture* and (3) *symbol*. Relating the notion of *intellectual material* to these modes, we see that it can be represented by each of them.

We state the following hypotheses:

A. Historically school has communicated its knowledge at level of representation (2) → (3).

B. Working class children as well as farmers' children usually possess knowledge at the level (1) and not at (2) and (3) levels.

C. In order to democratise education, as workers' children and farmers' children gain access to school, educationists should look for possibilities to include level (1), and to transform knowledge from this level into the two others.

The justification for (B) above lies in the strict division of labour which exists in industrialised countries. The point may be illustrated by an example from the building industry. Here certain people, such as architects, engineers, consultants, bureaucrats etc., plan and organise the building process. Other people, such as those preparing the ground, preparing the foundry, handling the cranes etc., perform the physical process. The division of labour, such as in this case, is manifested first of all in the principle that one group does not participate in the activities of the other.

Bernstein's (op. cit.) sociolinguistic concepts *restricted code* and *elaborated code* lend support here. The restricted code is the basis of the language of the working class, a language mainly related to concrete material. The elaborated code, located in the middle class, is the basis of a language which contains many more generalising and relational expressions.

The early Bernstein has been much criticised for describing the restricted as being inferior to the elaborated code. This view is no longer representative. It is now argued that working class language is as rich as middle class language: it is just the coding systems which are different (Rosen 1972).

I shall return to the importance of sociolinguistic research in §4.2. Here we note that sociolinguistics has already stated what I am arguing on behalf of folk mathematics: it is not a question of level, organisation or richness of knowledge; it is rather a question of how it is coded, represented, or worked out through activities.

Working class language and working class mathematics are as rich and rewarding materials for the pedagogue to exploit as any other. It is the activities in which language and mathematics are used which can differ from those of school.

Hypothesis C implies that manual work includes the use of mathematical knowledge. This assumption builds on the validity of the concept of Folk mathematics.

1.1.4. *The Advantages of Vocational Schools*

Such a theory immediately calls for several counter arguments. The first is that if one recommends a (1) → (2) → (3) model, why not the converse model (3) → (2) → (1) as well for children of white collar workers who, according to the theory, should have little knowledge at level (3)?

No problem. Such a situation would be ideal. The real problem, however, is that school has very few traditions of implementing the level (1) into its prevailing use at the other levels. When level (1) is practised, it is usually independent of the theoretical subjects represented at level (2) and (3).

Within the Norwegian scene, I have experienced that there is much to be learned from education in the vocational schools. Here, originally one and the same teacher would teach both theory and practice. This teacher would thus be in a position to genuinely integrate the two sides of knowledge. Take, for instance, the craft profession of *welding*. Here knowledge about geometrical shapes is indispensable. The properties of such shapes can be studied on the shopfloor by drawing on steel-plates, cutting them and bending them into cones, cylinders and more complicated products, for later review at the blackboard in the classroom. Alas, such a combination of theory and practice is dying out in Norway as these schools go comprehensive. Academic teachers are coming in and taking over the theoretical subjects and are not able to profit from the knowledge developed on the shopfloor.

The practical teachers have other stories to tell about pupils when they they start in the vocational school:

"It works as long as we do not call it mathematics. Often a boy will suddenly shout: Teach, you are cheating us. You called this theory, to me it looks like a damned piece of mathematics. And after this he would go on using 'terror(y)' for 'theory'."

A second counterargument to the theory sketched above is more difficult to handle. It is the argument that whatever the specifics of working class knowledge once might have been, they are no longer so important for education because modern times have levelled out possible differences.

Above all there is the fact that workers in the heavy industries today often earn more than teachers. And what about television and its impact on knowledge development? Such arguments bring up some of the basic theses of the theory in this book.

Knowledge and the development of knowledge are not solely dependent on fortuitous economic development over short periods. Knowledge, its forms and expressions, have a history which provide the individual with a basis for today's learning. The parents of today's children, even if they go to the Grand Canary twice a year, may have had seven years of unsuccessful schooling. Their knowledge culture, the way they store and transmit knowledge, their intellectual material, are not changed overnight by improved income.

As Leont'ev (1978) describes it when discussing the *activity* concept:

> man's psychological activity assumes social and historical structures and means transmitted to him by the people around him in the process of cooperative work in common with them.
>
> Ibid., p. 59.

When man's material environment changes, he still has to face it with the behavioural strategies he has coped with so far. What present and future transformations may lead to, because of changed environmental factors, is another question.

Underlying this reasoning is the hypothesis that the family is still the most important socialising agent of the child. This is much more questionable. There is much evidence to support the argument that the family's role in the socialisation of the child over the last decades has been reduced, at least in the Western world. We are thinking here of, for example, families breaking up because of changes in the employment structure (families having to split up or move), and the increased number of divorces.

Sociologists already recognise groups of children for whom the family has creased to function as a socialising factor, leaving this role more or less to peer groups. The implications of this for education in general, and mathematics education in particular, will be one of our concerns in later chapters, especially Chapter 5.

1.1.5. *Vygotsky*

Bruner has shown us various ways of preserving knowledge and demonstrating its existence in the individual. We still need theories about the relationships between the individual and knowledge. Bruner (1972) hinted at such problems, but he does not theorise about this.

Then Vygotsky arrived. His name was mentioned in various contexts placing him more and more in focus. He had been around some time, but we had not appreciated the depth of his work. He was the Soviet psychologist who said that external actions were internalised as thinking, and that thinking was thus structured as an inner language. He was given half pages in various American books on symbolism, language and cognition. And still only two modest books by Vygotsky are available in English.[4]

Pedagogues in nursery education were among those who demonstrated the power of Vygotsky's thinking. Their work on the relationships between play, work and cognition in the young child's life were very much based on Vygotsky's thoery of the function of symbolism in childrens' play, and its importance for later schooling.

Furthermore, language teachers started to use the term "functional language", by which they meant the use of language which functioned on the premises of the child rather than those of the school (or kindergarten); language was here a tool for expressing emotions, needs, claims, experiences, everything which was important in the child's life, rather than a tool for doing ready-made exercises in some textbook. It resembled the function I should like to see mathematics teaching fulfil.

Pedagogues in Denmark, Germany and Norway, restless and dissatisfied with the obvious stagnation of educational theory, started to investigate activity theory as a possible road along which to proceed. Activity theory has its roots in Soviet psychology, of which Vygotsky is recognised as one of the great founders. The theory is very general. It disregards major social structures which we today experience as important in order to understand learning processes within an educational system.

Culture is an example of such a structure. The role of oppression and resistance are others. I will examine these structures in the context of activity theory in Chapter 5.

1.2. THE FOUNDATION

> Literacy is not just the process of learning the skills of reading, writing and arithmetic, but a contribution to the liberation of man and to his full development. Thus conceived, literacy creates the conditions for the acquisition of a critical consciousness of the contradictions of society in which man lives and of its aims; it also stimulates initiative and his participation in the creation of projects capable of acting upon the world, of transforming it, and of defining the aims of an authentic human development.
>
> It should open the way to a mastery of techniques and human relations.
>
> Persepolis Declaration[1]

1.2.1. *Is Educational Activity Possible*?

The reader will from now on notice that I sometimes write "activity" with a capital A and sometimes not. One reason for this is that activity theory, without capital "A", as it was developed in Soviet psychology, lacks some concepts fundamental to an understanding of important features of Norwegian classrooms at least, and I will assume quite a few others. I shall thus make an attempt to analyse the possible foundations for a theory which includes, in particular, conceptions such as oppression, resistance and culture, as a basis for mathematics education. The theory I thus approach is Activity theory.

Activity is a way of describing the complete life of an individual. As an individual will always be considered in relation to the social groups to which he relates, Activity theory also describes the life of such groups. Activity will refer to actions emerging from the individual's own motivation. Activity is related to the individual as a *political* individual of society. This implies that the individual, as a member of society, is in a situation where he is permitted responsibility for his own life situation in particular and for society in general.

The strong component of "being an active member of society" in our concept of Activity implies that an individual can be denied access to Activity. This hypothesis will be explored in detail in Chapter 5. The problem we face is that we know about people being restricted to passivity in society, the living corpses or the silent masses which Freire describes (see §1.2.3.).

On the other hand, history is full of examples of groups of people who in the long run have combatted the oppressive forces denying them access to Activity. I shall argue, in Chapters 4 and 5 in particular, that the educationist has to relate her curriculum to the oppressive forces if she wants to promote Activities. Implicit in this statement is that failure to learn a school subject such as mathematics can be interpreted as the pupil's failure to experience the content matter as relevant to his Activities or to the oppressive forces denying him access to Activities. Furthermore Activity is related to the dialectic between an individual and his social environment. Thus, concepts such as *communication* and *culture* will be related to Activity.

Finally, Activity belongs to individuals. As educationists we can discover signals of the Activities of individual pupils. In accordance to these signals we can provide them with educational *tasks* which can develop and promote their Activities. As the above demonstrates, we have a long way to go. The rest of this book includes an exposition of the relationships mentioned above.

In order to develop the concept of Activity I shall start with Vygotsky.[2] It is so easy to underrate the significance of his use of this term, at least when reading the only two books by him published in English, as he employs "activity" throughout his writings in its common-sense way. Still "activity", as used by Vygotsky, labels what was to become the subject of influential developments within Soviet psychology.

It is mainly Leont'ev (1978, 1981), one of Vygotsky's successors, who treats Activity as a theoretical concept at the foundation of a scientific understanding of human behaviour.[3]

Both Vygotsky and Leont'ev build their psychology on Marx and Engels' conception of Man, as most Soviet psychologists would claim to do. The conception is one of man as an acting person, at one time being both determined by history and determining it, being both created by society and creating it. It is in this context of historical and dialectical materialism that Activity as an object for psychology is examined — as the process by which man acts in and on the world, transforms it and is transformed by it. The basic drive for Activity will be in production for survival: food, clothing, housing, etc.

The task for Soviet psychology here was immense. Concepts such as consciousness, meaning, motivation and many more had to find their place. Leont'ev (1981) points out that even after twenty-five years' hard work to develop the theory, many of its conceptions are still unsatisfactory and too abstract.

Of course, Activity theory easily becomes too abstract, and it may be difficult to see its relevance for education. Was it not precisely Piaget's notion, that the individual builds her intelligence by acting on the external world, constructing her knowledge as the result of her experiences?

And it is also very likely that other socialisation theories, such as Mead's, are more helpful for an understanding of the relationships between knowledge, meaning and motivation. I shall also explore the possibilities of Mead's theory in a later chapter (§4.1.). But Mead does not say very much about the impact of man's history on man's current activities; and he does not say much about the various thinking-tools and communicative tools for learning activities. Mead belongs to the large family of psychologists who disregard the importance of knowledge (such as mathematical knowledge in the form of thinking-tools) for man's development and potential in the world.

And it is precisely at this point that Activity theory offers itself as a generalising theory for the purpose of the educator who has some powerful knowledge, in our case mathematical knowledge, at her command.

Clearly we shall face many problems. I have already referred to such a grand term as "production for survival" as the basic motivation for Activity. One thing is that man does not collect berries and wood for the winter any more. At least, this is not the most important Activity for most people. A much more difficult problem facing the modern educator is that school, as the place where educational Activities have to be performed, is, as a result of history and the modes of production, a more or less closed system in relation to the rest of society. So when the foundations for an Activity are situated in history, production and society, we are right back to the familiar problem: how can we provide our pupils with experiences which can reduce school alienation?

Cole (1982—83), acting as a translator for Davydov, comments on this, quoting Davydov when he analysed the concept of activity: "But you'll never see educational activity in school." The quotation is reported here to demonstrate the difficulty of our project, not for the pessimism it contains.

The major obstacle may be, as Cole (ibid.) mentions, the bureaucracy of school. As I shall soon develop in more detail, the Actions comprising an Activity are goal directed, and it is vital for the individual that it is she herself who decides on these goals. As many of us know only too well, such decisions can be rare in many educational settings.

1.2.2. *Levels of Activity*

Activity theory, as we shall see, embodies the individual and society as a unity: the individual acts on her society at the same time as she becomes socialised to it. And for the purpose of the educationist, Activity theory has another great advantage: its key concept, Activity, focuses right away on what our project is usually about: the initiation of learning in the context of the classroom. Activity theory has the advantage for the educationist that it is a dialectical theory.

We shall not be studying learning solely in the context of the classroom. We shall also study learning outside it, and we will see how inside-classroom activities relate to outside activities. The dialectics here is located in the part — whole relationship: the classroom activities within learning activities as a totality which includes classroom learning.

The probability that topic A of a mathematics curriculum is recognised as important by a pupil is dependent on how he relates it to matters influencing his total life situation. On the one hand, previously learned topics of a curriculum will influence what is regarded as important in the context of the totality. A series of classroom experiences, over half a year say, will influence the pupil's evaluation of the next topic which appears at some level.

On the other hand, this new topic will influence the pupil's evaluation of the curriculum as he experiences this over time. Such a dialectical approach requires that the teacher (or the curriculum planner — I hope the teacher) thinks about the totality as well as the particular lesson. She has thus to relate the lesson to what embodies it: how does the lesson relate to her pupils' conception of the important totalities of their world, and how can this lesson eventually transform this totality?

It was Christiansen and Walther (1984) who first explored Activity as educational Activity, building a methodology and theory for the purposes of the mathematics educator. Building on the works of Davydov, Leont'ev and Markova, they analyse the relationships between educational tasks and activity in detail, demonstrating the power of the theory. I can only report some of the foundations they build on, as their goal differs from mine. They construct a detailed and deep educational theory about the various components of Activity. My goal is to look for relationships which prove significant as to whether an individual participates in the Activity intended by the educationist or not.

The problem here relates to the problem field which Bauersfeld (1979) recognises when distinguishing between the *matter meant*, the *matter taught* and the *matter learned*: the content matter of the mathematical structure communicated, the content of the teaching process as shaped by the teacher's learned structure and routines and the cognitive structure of the individual.

Bauersfeld describes it as an ideal case when these three forms coincide. From the position of Activity theory this statement has to be developed in terms of *context*: mathematical knowledge in its pure form as a structure of thinking tools is related to a system of contexts by the learner which defines the Activity the learner participates in. This system of contexts is dependent on the learner's history, such as her history within her culture, family, education and the subject.

Activity theory thus considers the relationships between the content matter of a learning situation and its context as the latter is defined by the learner, and the dialectics between these two, that is, how they mutually influence and develop each other. This dialectics will be examined further in §4.3.

Bauersfeld guides us on to an exciting and important track: the search for correspondence between the teacher's expectations of her pupils' learning and the pupils' own motivations for various sorts of knowledge. To get somewhat closer to the concept of Activity we should note that it can be examined at various levels (Leon'tev 1981; Wertsch 1981).

Activities cannot be examined without recognition of their motives and the object towards which they are oriented. Motives are decided upon and determined by the individual (usually in cooperation with other individuals; Activity is always social). An Activity builds on

Actions, which are connected with their goals. It is still the individual who decides and determines.

In the words of Christiansen and Walther:

> The relationships between, on the one hand, *motive and goal*, and, on the other hand, *activity and action* may thus briefly be described in this way: the flow of a given internal or external process of activity/action develops and proceeds *with respect to the motive* (the factual object) as activity, *and* with respect to the goal (*respectively the system of goals*) as action.
>
> Ibid., p. 37

The whole point for the educator to recognise now, and to take advantage of, is that whatever she observes of learning behaviour by her pupils, this behaviour is part of some Activity, and she has to learn what this Activity is about in order to create a constructive encounter between this Activity and the various educational tasks she can provide. The problem is to know about *which* Activity the learner will relate to the educational situation with which he is confronted.

This point is easy to understand when applied to incidents outside school. If I observe (a) some young men getting into bathing suits on a cold Winter day, (b) jumping into a river of not very clean water, (c) apparently having great fun although they are slowly getting blue, (d) being accompanied by cheers from the crowd around, I shall probably get some funny ideas about the people I am observing. The various Acts (a-d) do not provide me with much meaning or understanding so far, although I can admire some of the behaviours I observe (nice jumps into the water, a good crawl etc.).

A minute later, however, I observe (e) some of the people around collecting money for the show, and finally (f) handing it over to a representative of some charity fund. By then I have understood what it all (a-f) is about. I have learned something about their total Activity.

The same sort of situation occurs when we observe children's play. We can see children build with their construction toys, such as Lego. We see them build (a), build (b), and so forth, and we observe a series of Actions, without necessarily knowing what their Activity is about. Only when I see the child put the whole lot together to make a fancy, grand aeroplane do I understand what she was doing.

It is the sort of situation which occurs when we read a novel where the author is slow to show her cards, or watch a film where the build-up is equally slow. We can gradually discover the artist's project, by experiencing the various *segments* of it. The situation in the classroom is much more difficult. We observe each pupil in short glimpses; we observe the segments of their Activities; and often we do not manage to identify their Activities at all. In other words, we do not discover what the real objectives of our pupils, as related to school learning, are about. Bateson here would have stressed that one cannot understand learning behaviour without knowing the context that the individual learner relates to her learning.

Activities are thus about the decisions, projects and corresponding goals of the individual. As a teacher I can only observe them and make an attempt to understand what they are about. I can evaluate them, saying that some are destructive (as in the case of violent behaviour); and, not least, I can provide my pupils with situations intended to initiate constructive Activities.

It is in connection with this that Christiansen and Walther make their analysis, discussing the dialectics between Activity and educational tasks (they do not use a capital "A"). The educational task is set by the teacher. The tasks of mathematics education are all those familiar components of our lessons: problem solving, routine exercises from the textbook, the learning of a mathematical principle, applying mathematics, environmental projects etc. Such tasks relate to the learner's general Activity and specific educational Activity: "The task and the activities establish so to say the "meeting place" between teacher and learner" (Ibid., p. 8).

The role of tasks, still according to Christiansen and Walther, can thus be considered on two planes:

- the activity of pupils can be initiated by means of tasks;
- to motivate for specific types of activity, such as *exploratory activity* or *problem solving activity*, specific tasks are needed.

Christiansen and Walther proceed by examining the role of the learner's *regulation* of activity as related to the meeting place mentioned above. The significance of such an analysis can be recognised in the perspective of the recent developments within mathematics education described in §1.1.1., referred to there as the principle of *self-activity*.

It is such regulation that I have interpreted as a dialectic above: the Activity is related to the totality and the tasks to the parts of the totality. Tasks are in the hands of the teacher, and her success is dependent on her insight into the Activities of her pupils.

1.2.3. *Activity is a Political Concept*

In its broad sense Activity is a *political* concept. This too is the reason for the title of this book. As far as I can discover the Soviet psychologists never cared to stress this point. Again, we have to consider the kind of historical context in which they developed their theories. In societies where there are conflicting ideas about what the important objectives for life are, and about what the goals of society should be, and so forth, the political component of Activity can be analysed at various levels.

It is the macrolevel which I shall return to repeatedly throughout this book. At this macrolevel we shall discover that groups of individuals, even nations, are prohibited from Activity, being in an oppressed political position. This is what Paulo Freire's works are about.

Subordinate levels can be recognised in relation to motives and goals. Activities represent how a particular individual decides to act in her world, according to the make-up of this world. Individuals do not always agree on which Activities are the important ones to carry through, or how to carry out any particular Activity for which the goal is agreed. Thus, I shall discuss the concept of *ideology* and its relation to Activity in Chapters 4 and 5.

One of the difficulties arising from such a project is that when ideology has been examined in the educational context, it has usually been related to school as an institution or as part of the State Apparatus (that is, school as a producer of ideology). In the context of Activity theory, we shall have to look at the pupils themselves as carriers of ideologies, and the implications of these ideologies for our curriculum planning.

A gang of youths can have an intentional politics of hooliganism as a way of coping with life-at-the-moment. They can also have a politics of turning their backs on school. Their Activities will accord with their politics. Later we shall see quite a few examples of young peoples' Activities, in which mathematics plays an important role in the achievement of political goals. Just to give a hint of what such goals can be about: questions concerning the improvement of road safety, drug safety and the size of the youth clubs nearby.

The reader should thus have been warned: the use of *politics* in this book does not refer to any conception of politics which has to do with her vote every second or fourth year. It is not related to a Labour, Liberal or Conservative ideology in particular. And it must by no means be confused with *indoctrination*, which implies possible oppression of specific ways of Acting.

The term "politics", as used here, is related to the position that human beings act, participate and survive in their world as *political human beings*. It is a variant of the "man is social animal" thesis. Men not only act together with other men in social settings, they think differently about important matters in their lives as well, thus being, by nature, political men.

1.2.4. *Activity Is Social*

In the broad sense Activity is the way Man acts in his world, transforms it, and is being transformed himself in a variety of ways. Such transformation takes place in environments which are primarily social. According to Soviet psychology, Activity is social Activity. There is no place for Robinson Crusoes here.

> In all its distinctions the activity of the human individual represents a system included in the system of relationships of society.
>
> Outside these relationships human activity simply does not exist. Just how it exists . . . cannot be realized otherwise than in the concrete activity of man.
>
> Leont'ev 1978, p. 51

This is almost the same as Bateson's (1973) claim that mind is part of a greater Mind, which is immanent in the environment of the individual.

One of the problems the educationist faces here is to determine where the boundaries of the significant environment are located. One can go the whole way, seeing how the pupil is under the influence of her family which is under the influence of employers who are under the influence of the government, and so on. There is always a way of explaining behaviour in this sense. On the other hand, educational research has often recognised too narrow boundaries, usually those of the classroom, thus offering little help for teachers who face pupils who are strongly influenced by factors outside the classroom.

The boundary may be located somewhere between the classroom and Parliament. In most cases connected with the relationship between the pupil and her school, we should probably do wrong to depart too far from the school. The various examples of projects provided throughout this book will give some indication about where the educational Mind can be found.

It is interesting to see how Freire's (1972, 1975) concept of man comes close to Vygotsky's and Leont'ev's. Freire developed his theory about *conscientisation* 30 years after the works of Vygotsky appeared, and yet he apparently did not know of them. He developed his theoretical framework in a country where most of the population suffered heavy oppression. Freire distinguishes between man's *integration* and *adaptation* to society, and relates these two processes to, respectively, a *subject* or an *object* in society.

To the extent that man is not prevented from acting in his world, not prevented from choosing and making decisions about his actions, he is in a position where he can both adapt to society and transform this reality. He is then integrated into society. If not, he is subjected to the choices of others: his decisions do not belong to himself but result from external prescriptions. In this case he is only adapted to society, being objectified in it.

For Freire, contrary to Soviet psychology, it is important to conceptualise not only Activity, but also the restrictions which can be made on Activities, resulting in passivity, silence, and distortions of behaviour.

Following my introduction the reader can perhaps grasp a picture of one group of pupils for whom mathematics, as usually experienced in school, has such significance for their life Activities that they learn it. Furthermore we can perhaps see the picture of another group of pupils emerge, for whom mathematics education, as they have usually experienced it, is not recognised in this way. Because of this, pupils of the latter group may gradually turn their back on mathematics, ceasing to learn the subject.

It is the relationships between mathematical knowledge as experienced by the pupils and their possible Activities which build the foundation for a politics of mathematics education. The strength of Activity theory, as compared with other educational theories, is that it unifies society and the individual. The history of social science demonstrates how needed such a theory is. On the one hand we have a wide range of theories of social reproduction which see man as being determined totally by the reproductive forces of society. (I shall return to some of these in §5.1.2.) On the other hand we have a social psychology which neglects any influence of society on behaviour. Activity theory stresses that the individual acts within social structures, and thus both creates these and is created:

> if we removed human activity from the system of social relationships and social life, it would not exist and would have no structure. With all its varied forms, the human individual's activity is a system in the system of social relations. It does not exist without these relations. The specific form in which it exists is determined by the form and means of material and mental social interaction (Verkehr) that are created by the development of production and that cannot be realized in any way other than in the activity of concrete people.
>
> Leont'ev 1981, p. 47

The English social theorist, Anthony Giddens, has built his own theory of social actions. He does not refer to Soviet theory, but his paradigm is the same as the one stated above. He has conceptualised the notion of *structuration* (Giddens 1976, 1977, 1979). This refers to a theoretical position in which the individual and her social structure are considered as having mutual influence on each other:

> By the duality of structure I mean that the structural properties of social systems are both the medium and the outcome of the practices that constitute the system. The theory of structuration, thus formulated, rejects any differentiation of synchrony and dichrony in statics and dynamics.
>
> Structure is not identical to its constraints.
>
> . . .
>
> Structure is not to be conceptualized as a barrier to action, but as essentially involved in its production.
>
> Giddens 1979, pp. 69—70

The importance of such a conception is its duality for actions: at the same time as the individual lives within certain controlling structures, she creates these structures by her activities.

Giddens thus gives us clues about how to think dialectically about the possibilities for man of Activities within given structures. Having done so, Giddens (ibid.) more or less turns his back on the future, and instead chooses to examine the history of social thinking by means of his thinking tool. As a result we miss the analysis of Activity, or social action in Giddens's terms, as a drive for history or for social development. This becomes most obvious with Giddens's conception of ideology, which appears independent of Activity in his theory. I shall return to this in §5.1.

1.2.5. *Communication Is Part of Activity*

The social element of Activity is demonstrated in the conception of *communication* within the theory. The Soviet psychologists do not agree about the relationship between communication and Activity. Enerstvedt (1982) examines the contradiction between Leont'ev and Lomow on this problem.[4] In what follows I shall adopt Leont'ev's position.

The issue is whether communication is part of an Activity or not. This problem is connected to the issue of who is in a position to perform an Activity. Is it an individual or is it an individual as part of a group? If it is mainly the group which is behind an Activity, communication will necessarily be a part of it. This is the view of Leont'ev. He defines communication as a system of goal-directed and motivated processes which ensures the interpersonal components of Activity.

It is the group, or the collective subject, which is the centre of an Activity. The individual exists only in terms of the group. If the individual performs some tasks by herself (such as doing some mathematical problem), she still does this in relation to the group to which she belongs if her problem-solving is to obtain status as Activity.

Communication thus becomes an indissoluble part of the Activity process. It is through communication that ideas are shared, strategies developed, and projects carried out. The view of Lomow on this is that Activity can also be a subject-object relationship such that Activity can relate to a single individual. Communication will accordingly be a subject-subject relationship, which can be *about* Activity, not necessarily a part of it or a vehicle for it. I adopt Leont'ev view that the group is the subject of an Activity: that is, it is the group which acts it out. Individual behaviour is to be interpreted only in relation to collective behaviour, that is, group behaviour. It is in such a context of group behaviour that communication is to be understood and examined.

This secondary school was only a few kilometres from Bergen Airport. It was the last one I worked in as a full-time teacher. We discovered well-hidden plans for extension of the runway. It turned out that the extension would go through some of the families' gardens. The noise around the homes of several of the pupils would be earsplitting. We wondered about the cows and the horses. Should they all wear ear mufflers? What would happen at the workplaces? Would there be more jobs or not?

Our school was an idyll in the countryside. And five minutes away someone planned to develop an airport that would be the major transantlantic port for Scandinavia.

We constructed a project. Its major object (i.e. what would be an educational Activity for several of the pupils) was to inform parents about the problem. Very few knew. We collected information. We coloured agricultural maps. On other maps we coloured the new noise zones, counting how many people would live within the worst zones. Some of the pupils built a topographic model of the extension. Some interviewed the director of the airport. Some interviewed the politicians and discovered that many of them knew nothing. We wrote a white paper for distribution and we made an exhibition.

The pupils worked individually and in groups. But they all worked according to their ability and interests in the context of the project, the research. Some mathematics was used as thinking tools: units of area, ratios, logarithms (decibel formulae for the noise zones, official definitions) etc.

"Ability" in this project was related to its contribution to the collective effort, not to any ranking system for the purpose of ordering the pupils according to some scale. It is such situations, such projects I shall be looking for. The problem, which occurred here too after four weeks, is the schizoid situation with which we have to cope in societies based on individualism and competition. It was Anne who came up to me, saying:

"Teach, what we have been doing so far is all right, but you see, I am going to be a nurse, and the other class is half-way through the book by now. Are we going for long with this project?"

This problem, familiar to everyone who has ever been inside a classroom, is both practical and theoretical, and accompanies most of what we are doing within education.

Giroux (1981) points to what he sees as a weakness with Freire's theory here. Freire does not analyse the relationship between Activity and communication sufficiently. Freire repeatedly stresses that theory relates to Action, and conversely. But his conception of conscientisation (critical awareness) as he practises it, does not include Activity; it is only taken as a necessary condition for Activity. Thus conscientisation (to be described in detail as practised by Freire in §5.2.), becomes a communication process divorced from Activity.

The remedy for this is obviously to keep trying to catch the future in the present, or to reflect on Activity as a future project in the present whenever communication is reflected upon. The position that I take in restricting my conception of Activity to collective behaviour may be difficult for my fellow educationists to accept. Most of us have been trained as educators just to foster *individual* performance, and most educational theories are theories about individuals as well.

Pupils of modern Western societies learn within competitive educational systems, which have both selective and controlling functions. However, to say that pupils in such systems, based on individual performance, will be prevented from Activities because of the nature of the system would be to exaggerate.

First of all, there will be room in most schools for tasks other than those fostering individualism. Secondly, if our conception of Activity is going to have any relevance at all, educational Activities such as learning Activities within mathematics must contribute to the stock of knowledge of the individual pupil in such a way that it improves her individual performance. Individual performance as related to the school's sorting and controlling function, is irrelevant to Activity theory. As a learning theory, Activity theory does not include individual goals.

Finally, another strength of Activity theory is that it can challenge individualistic praxis,[5] as it can guide a group of pupils and their teachers say to increase their potential field for collective learning. More simply, pupils and teachers working within a tradition based on individualism can challenge this situation by insisting on the introduction of more projects based on cooperation, that is, possible Activities.

It is precisely this that Giddens (op. cit.) describes as *structuration*: individuals have to act within social structures, at the same time as they can create these structures. As we will see later (§4.3.), there is much that suggests that such a transformation of education from individualism to collectivism, or from individual learning to Activities, will be a key factor for improving remedial education. Nor can it be said that Western models of societies and education will be predominant in tomorrow's world. It is no coincidence that we can learn so much in the West from the Brazilian educator Paulo Freire, who has developed his methods in a large number of African and South American states.

1.2.6. *Internalisation*

I hope my reader by now has some picture of an Activity as a process in which the individual thinks and behaves, but does this in relation to a larger project and in relation to a group. Obviously there is some connection between the thinking of the individual and the group's thinking, and the other way round.

It is the theoretical description of such a dialectic which is one of Vygotsky's great achievements in psychology. Calling the communicative Activity within the group *Interpersonal* Activity, Vygotsky investigates the internalisation of this, and names the resulting "inner" Activity *intrapersonal* Activity.

Internalisation is the internal reconstruction of external behaviour. The vehicle for these transformations is primarily language. It is in this field of internalisation that Vygotsky did most of his research on children's cognitive development, a research I shall profit from in the next chapter on mathematics as a language.

Before I proceed to describe the major theses of Vygotsky's theory, we can now see how *meaning* finds its place. The meaning of some knowledge is developed jointly with an Activity. It has thus two dimensions: the shared dimension through interpersonal Activity, and the subjective dimension as represented by intrapersonal Activity. The first dimension is the objective dimension, the latter the subjective. As meaning exists in the form of language, language is shared socially as an objective reality. The meaning which language conveys, however, is interpreted subjectively by the individual.

Leont'ev expresses this in the following way:

> Activity enters into the subject matter of psychology not in its own special "place" or "element" but through its specific function.
>
> This is a function of entrusting the subject to an objective reality and transforming this reality into the form of subjectivity.
>
> Leont'ev, op. cit., p. 56

There is a shared, for the individual external, Activity, and a meaning connected with this. But the individual relates herself to this Activity also by her subjective interpretation of it.

We now face Mead's theory, and Blumer's conception of a *joint action*. In §4.1. I will refer to an example of a joint action, *dinner*. A dinner can be analysed in terms of the inter/intrapersonal dimension. We find here the joint actions (Cheers everyone) which can be interpreted subjectively (That was the fourth in five minutes), and where the mutual influences on the total Activity, the dinner, can easily be traced.

Freire lends support here as well. To Freire *meaning* is *shared* meaning. His stress on dialogue demonstrates this: meaning develops between people, not in people; knowledge does not exist in people in their world, but *with* people *with* their world.

Going back to the discussion in §1.1.2. about *mathematics* as related to *folk mathematics*, we see that mathematics can be considered as interpersonal knowledge. We experience discussions in the classroom based on some intellectual material as material connected to folk mathematics. Such intellectual material can either be intrapersonal, i.e. subjective, or interpersonal, i.e. objective.

The collective exploration of such intellectual material can lead to some shared knowledge, interpersonal knowledge, mathematics.

Fig. 1.2.1. The eight-leaf rose.

I can give my students the above diagram which, in Norwegian female culture, is known as the "eight-leaf rose". It is a familiar pattern in embroidery and knitting. I can ask them to draw the rose and be aware of how they count at the same time.

Each student has their own way of counting based on various conceptions of the symmetry of the rose. To the extent that such counting systems have been passed from mother to daughter, the intellectual material they represent is shared, interpersonal knowledge.

Individual, or personal subjective systems of counting will be intra-personal knowledge. The feature here is that the intellectual material, either shared or individual, is the basis for a discussion of mathematical ideas, in this case the analysis of various concepts of symmetry. Following this, the mathematical concepts of symmetry can be developed as interpersonal knowledge, and possibly function as shared thinking tools in the future.

The notion of mathematics as shared and interpersonal knowledge raises an issue which I will analyse further in Chapter 5: Who is it who is going to share some knowledge as mathematics? Is it the pupils of a school class? Is it the members of a culture or a country? A social class?

The issue has also been made pertinent as anthropologists reveal how different cultures build different algebras and geometries (Chapter 3). The notion of mathematics knowledge as interpersonal knowledge is thus a dialectical and political conception, as I shall develop further in later chapters.

This notion of mathematics is dialectical: On the one hand there are the pupils' discussions about a particular idea with the potential to develop into a theoretical and mathematical idea; on the other there is what has acquired status through the official curriculum as *the* ideas. This is the same sort of dialectic which is present in all processes of socialisation: on the one hand the desire to socialise the individual to the norms and standards of society; on the other, the ideology which speaks of the creative and free citizen.

Christiansen (1984) points to this dialectic as a *potential*:

> The relationships between *personal, shared*, and *objectified* knowledge provide rich dialectical potentials for the construction of knowledge of each kind. These dialectical potentials are due to the distinction and differences *within* each domain and *between* the domains. Thus, for example: the differences from person to person caused by the specificity of personal knowledge and experience; the differences between the individual's knowledge and that shared by the group; the differences between various conceptions of the knowledge "shared at present within the group"; the differences between shared knowledge within the group and knowledge objectified externally to the group; the differences between various representations of objectified knowledge.
>
> Ibid., p. 148

He then proceeds to stress how the internal interplay between an individual's store of knowledge and his activity/actions "may serve as a powerful source in his development of new personal knowledge and new potentials for activity/actions". Christiansen furthermore makes the same consideration for the group: "the interplay between the store of knowledge inherent in a group and activity/actions performed by the group may serve as a powerful source in the development of personal, shared, and objectified knowledge".

The notion of mathematics is also political. The hegemony of certain conceptions in mathematics is standard: we examine the properties of some intellectual material, we see their potential for developing into mathematical ideas, leading to one particular mathematisation. The politics of mathematics is here related to the "democracy of ideas" to the extent that a group of pupils is permitted to build its own theoretical ideas on the basis of some material.

1.2.7. *Tools*

It is primarily the role of *tools* which makes the mathematics educator so pleased about Activity theory. Perhaps for the first time ever we are offered an educational theory which considers what most of us are concerned about: mathematics as a field of knowledge comprising powerful knowledge. Many of us have intuitively felt that *mathematics* as a field of knowledge is important knowledge, not only for the benefit of industry and technology, but also for the individual pupil. We have, however, had little to offer when meeting the soft educators of the "I-teach-children-not-subjects" type. In other words we have been able to argue for the importance of various mathematical concepts and algorithms. Our problems have been mostly theoretical, to embed mathematical knowledge into an educational theory which considers the social and psychological aspects of the pupil. It is here that the Soviet psychologists make one of their greatest contributions recognising the importance of *knowledge* for the political human being. They do it thoroughly, examining man's use of tools in both ontogeny and phylogeny.[6]

We usually think about tools as items external to our bodies: hammers, rakes, spades. Such items are still tools in Activity theory. But the concept is generalised to include *thinking tools* and *communicative tools.*

It is the study of *tool use*, including thinking-tools and communicative tools, which is another of Vygotsky's great achievements for educational psychology. And it is the under-estimation of *Activity* in connection with such use which is one major reason why the strength of his theory has not been recognised fully: *the development of tool use is related to Activities.*

In this sense we can regard *toys* as tools and *play* as Activity. The child's sign systems will be its communicative tools. It is the relationship between the various uses of tools which lies at the centre of Vygotsky's theory.

Parallel with the development of the use of external tools in the child's life (as in the use of toys), speech develops. The child is responding to its social environment by adopting its language.

It is a result of phylogenesis that the child, unlike nonhuman beings, is equipped with prerequisites for speech. But the child has to *learn* to speak, the development of speech is dependent on her social environment. Speech is a sign system. A gesture system such as those employed by children in their play, is the same. Written systems are sign systems: written language, written mathematics, written music, Braille and so forth. Sign systems are communicative systems, developed for their different purposes, and they belong as tools to Activities. Speech, however, holds a predominant position among the sign systems. It is the one most human beings meet first, it is the one used mostly for communication, and it is the one on which the other systems build and rely. Speech thus guides, determines and dominates the course of Activity, enabling the child to master her environment.

But gradually the child is no longer dependent on audible speech; she can solve her problems by speaking to herself, with inner speech. It is here that the contradiction between Piaget and Vygotsky becomes clear. They both describe *egocentric* speech. For Vygotsky this is audible speech before the child masters its internalisation. Such use of speech is still intelligent, as a constitutive part of the child's Activities.

For Piaget egocentric speech is mainly proper to a particular stage in the genesis of the coordination of actions: the child is not yet able to reverse her action and thus perform operational thinking. The development of speech, particularly non-egocentric speech, will follow in due course.

I would again stress the intentions behind the theories: the biologist Piaget aimed at building a complete theory of human development in terms of the growth of intelligence. Vygotsky, originally a lawyer and philologist, set out to build a psychology at a time his nation had acquired a socialist state, probably one of the most dramatic moments in history. Vygotsky's psychology had necessarily to have the history of man as one of its cornerstones, in order to create the new Man, the Man in his new State as a social man.

Furthermore, the majority of the population of the young Soviet state was illiterate. *Education* thus became indispensable. In the end Vygotsky also built a psychology for education, his *pedology*. Returning to mathematics education, we can regard *mathematics* as providing tools, both thinking-tools and communicative tools. Their use is related to Activities. Their functionality is dependent on whether they are experienced in the process of Activity or not (see §1.2.10). Speech has a predominant role as a tool. It is the first thinking-tool and communicative tool the child learns to master and which will be the basis of future tools of this sort.

The concept of *possession* of a thinking-tool is related to the teacher and learner. Consider the current practice in the classroom. We teach the build-up and the uses of a thinking-tool. In former days we talked about the method of example-and-rule. After Piaget we would say: Provide the pupil with some material so he can experience the concepts and structures, and eventually go on to routinise knowledge.

Who possesses the knowledge? Who poses the problems? Is it a sufficient condition for the possession of the pupil of the thinking-tool that he can use it appropriately according to teacher and textbook standards in situations designed by the same?

Sometimes perhaps. But consider the following situation.

The teacher communicates the uses of some thinking-tools, some parameters of descriptive statistics, say, the arithmetic mean, the median, the mode, maximum and minimum of a set of numbers. He checks that all the pupils know how to use these tools. So some empirical set of numbers involving more than one variable is presented to the pupils.

"Use the tools you have learned. Investigate. Describe."

This situation differs from most educational situations where mathematics is involved in the sense that the pupils have to decide how to use the tools. Empirical material involving several variables can be structured in different ways, and the tools can be used in various ways. Each choice leads to a specific conception of the material and is based on certain goals and motivations in the pupils.

The significant feature here is that these choices of structure and tool are up to the pupils. The teacher does not tell them how to investigate and what to investigate. She leaves the investigation the her pupils. The argument is that in these cases the pupils will *possess* the thinking-tools in different ways. In one case it is the teacher who directs their uses, in the other the pupils themselves.

We can go further with this example. In the above project it was the teacher who presented the empirical material. Consider the following ideal situation: the pupils discuss some matter. They disagree. Argument opposes argument.

"Let's investigate, so we can see how valid the arguments are."

"How?"

"Let's use what we learned in statistics!"

In this case it is the pupils who define the project in which the tools are to be used.

It is hard to see this situation occurring inside the classroom, as most activities there are directed by the teacher. Perhaps we can imagine the teacher provoking her pupils when she listens to such a discussion, asking them to investigate.

Again it is a question about *possession* of her thinking tools. The concept of *possession* is related to Activity. As we rarely know for sure what the Activities of our pupils are, we will rarely know much about to what extent they possess the thinking tools of mathematics. By their behaviour we observe the signs and interpret them.

1.2.8. *The Role of Speech*

As we have seen, it is the dialectic between the emergence of tool Activity and speech Activity which is Vygotsky's interest when he studies the development of intelligence:

> The most significant moment in the course of intellectual development, which gives birth to the purely human forms of practical and abstract intelligence, occurs when speech and practical activity, two previously completely independent lines of development, converge.
>
> Vygotsky 1979, p. 24

There is no form of isomorphism between the two systems, in the sense that an incident of tool use can be mapped onto a corresponding speech act and conversely. It cannot be so, as the two systems have different origins both in phylogeny and ontogeny. The two systems were originally used for different purposes: the use of tools was directed towards nature, the use of speech is a social communicative act.

The child learns to say the names of the tools, the names of the actions they perform, as sound images, which bit by bit are internalised as meanings. It is this growing role of speech as a tool for the child's Activities, which is Vygotsky's trump card. At one stage

> *children solve practical tasks with the help of their speech, as well as their eyes and hands.*
>
> This unity of perception, speech and action, which ultimately produces internalisation of the visual field, constitutes the central subject matter for any analysis of the origin of uniquely human forms of behaviour.
>
> Vygotsky, ibid., p. 26

Vygotsky can now conclude that children solve their practical tasks with help of their speech, as well as their eyes and hands. He stresses how such use of speech increases the child's freedom to act, her spontaneity, and how the child's actions are also controlled by speech behaviour. The next dramatic moment in speech development comes when the child internalises language behaviour, when she is thus no longer dependent on audible speech:

> Instead of appealing to the adult, children appeal to themselves, language thus takes on an *intrapersonal function* in addition to its *interpersonal* use.
>
> Vygotsky, ibid., p. 27

The child now speaks to herself by inner speech: she can reason without saying her thinking out loud. It is this moment in the child's life which Piaget also signifies and describes as equipping the child with a new logic, characterising the stage of concrete operations.

We note the term "function" in the above quotation. We shall meet it repeatedly from now on. It is what we centre on: to examine conditions in which knowledge can be functional for the pupils. In other words: to develop thinking tools in the context of Activities.

1.2.9. *The Discovery of a Thinking-tool*

In 1981 I did some empirical research on teachers' conceptions of the "practical" in relation to a mathematical problem (Mellin-Olsen 1981). Synonymous with "practical" the teachers were offered with "being relevant to everyday experience". After some pilot investigations which revealed that the teachers' conceptions were quite diverse, I designed the topic as the theme for an in-service day at three infant schools. I also tested my student teachers who followed our 1/4 year course in mathematics education. After introducing what it was all about (We are going to analyse the concept of practical mathematics, relevant to everyday life, in the context of a textbook problem), I confronted the teachers and the students with a sample of 36 textbook problems, which they were to evaluate as practical or not. A discussion followed.

> Carl's largest carrot is 12 cm.
> Else's biggest carrot is 2.0 dm (decimeter).
> Which carrot is the largest?

Teachers' responses:

- This problem is practical since it requires practical knowledge to compare length measures.
- The problem is not practical since it involves the use of dm which is not much used in daily life.

- The problem is not practical since it is not likely that the kids ever will compare lengths of carrots by cm or dm. This problem would have a chance if "fish" was substituted for "carrot".
- The problem is not practical since it is hard to see the purpose of the context of the problem.

Erik divides the way to school into three parts.
1. 235 m from home to Hansen's supermarket.
2. 348 m from the supermarket to the bus-stop.
3. 171 m from the bus-stop to the school

How long is his way to school.

Two concepts of a practical problem emerged:
- The problem is practical, as it can be interesting to know about how far it is to school, and this problem demonstrates how it can be done.
- The problem is not relevant, as the algorithm used (division into three parts) does not seem to be efficient for practical purposes.

Following such discussions some more precise formulations emerged:
- A problem is practical if it has a non-mathematical context.
- There is present a relevant *purpose* in a problem. The purpose has to be demonstrated through formulation of the problem.

This was developed further by one student;
- A practical problem is a problem which the learner will with a certain probability face and which he will want to solve.

An interesting outcome was that the students did not accept as many problems as being practical as the teachers did. 28 students evaluated 33.8% of the sample of 36 problems as "practical". The corresponding figures for the teachers were 57.2% and 49. Some of the difference in these figures can be explained by the students' weekly didactical explorations into such matters, which were intended to provide them with a theoretical stance in relation to this task, whereas the teachers' daily educational activities are related to texbooks which are packed with problems of doubtful content.

Activity theory will test such problems by the strongest criteria of them all. "Likely to meet outside the classroom" and "purpose" seem to be two good standards to set for such problems. Burkhardt (1983) analyses similar questions by introducing the concepts *action problems* and *believable problems.*

It is important to note the individual's *discovery* of the functionality of a sign system, or a thinking-tool. One of the most impressive cases of such a discovery is the story of Hellen Keller. She was born deaf-blind. At the age of 10 she first discovered the *function* of the signs her teacher wrote in her hand. It was the sign for "water" that finally led her to this discovery. Before this her behaviour could best be described as being like that of a well behaved animal, without any developed sign system as a communicative tool or thinking-tool.

Land and Bishop (1966—69) demonstrate the principle of discovery in connection with pupils' use of graphs as tools for problem-solving. They discovered that some pupils would invent suitable graphs in order to solve particular sets of problems. The researchers returned to their classes and drew the attention of the rest of the pupils to the usefulness of graphs. On retesting, it turned out that the pupils who originally knew about the graphs as a functional tool could invent new graphs as tools for new problems, while the other pupils could still not exploit graphs in this way.

The findings of Land and Bishop point to two extremely important phenomena for mathematics education, which I shall repeatedly return to:

(i) The discovery of the functionality of a sign system by the individual. This discovery is partly the result of being familiar with the use of the sign system. It is dependent on the presence of an Activity.

(ii) The importance of the presence of a decision by the individual to employ (or not to employ) the sign system. This decision is a social decision, made by the individual in relation to others (see §4.1.).

(ii) above assumes that it is not sufficient to consider whether a pupil realises that the option represented by the new sign system is available or not, or whether she knew how to use it or not. The statement relates to the relationship of possession of knowledge: the learner will evaluate the new knowledge as she learns it, and this evaluation will have influence on her further learning, i.e. her decisions on her own learning strategies. This process is social, the decisions made by the learner are made in relation to others (see §4.1.)

It is such discoveries and decisions that most mathematics education is about; the goal will usually be to make them happen. We notice the flavour of Piaget all the way here: we cannot *persuade* children to learn, we cannot *demonstrate* basic knowledge to them (or general operations in Piaget's terminology), expecting the children to generalise the knowledge. The only thing we can do is to provide the children with a sufficiently rich learning environment, in which they can play, experiment, experience and so forth, for their own construction of knowledge. Translated into Vygotskian theory, this means that no one can convince or persuade a pupil of the power of a language or other thinking-tool: this discovery depends totally on her experience of its functionality connected to something she evaluates as Activities.

1.2.10. *On Functional Literacy*

I shall make a detour to analyse functional literacy. This may be well worth doing as it will give us some flavour of what the *functionality* of some thinking-tool or communicative tool can imply. Furthermore *arithmetical skills* are often included in recent literacy programmes.

The notion "functional" is a typical case where a word or a label arises from an ideological struggle. UNESCO especially has been criticised for reducing functional literacy to the trivial use of the written and spoken language (Levine 1982; Mackie 1980). This criticism has indirectly been met, as we will soon see, by several UNESCO officials.

The banalisation of the term "functionality" can be observed in cases like textbooks, such as *Penguin Functional English. First Impact.* Here the first unit on page 8 reads:

> Brian: Hello Janet. Nice to see you. How are you?
> Janet: Fine, thanks, Brian. And you?

Such an opening does not necessarily reduce language use just to trivialities. But *Penguin Functional English* continues to trace the same pattern that it laid out at the beginning, a sort of pattern which Freire (1970) also illustrates from an adult literacy program:
"Charle's father's name is Antonio. Charles is a good, well behaved, and studious boy." Freire goes straight on hammering, describing such content of a literacy program as being deprived of reality, "thus being impoverished, they are not authentic expressions of the world."[7] It should not be necessary to draw parallels in mathematics education.

"Antonio has three apples. Then he gets five apples."

Or: "Carl's carrot is 12 cm and Elsa's is 2.0 dm. Which one is the largest?"

It is an old story, and we have heard it too often. So how are we going to challenge such an "impoverished" conception of functionality?

Remember now that according to Activity theory functionality is about a relationship between the individual and her Activities, being a property of her thinking-tools or communicative tools. Furthermore there is an element of discovery and decision by the individual in relation to this property. Many of my readers can now argue that they, as I myself, were victims of a "how are you" literacy campaign when we started school some decades ago. And still we discovered functionality. We can write essays, petitions, even books. But that was us. After the approach we were exposed to, both in language and mathematics education, many more did not acquire the relationship to thinking-tools and communicative tools that we did.

Our thesis, already stated several times, is that in order to discover functionality, the individual has to relate the tools to some projects of her own. Obviously for several pupils school in itself will be such a project (as I shall discuss in §4.1.), but equally obviously, independent of its content for other pupils, school will not be such a project. Following this we will have to pay attention to the content-matter of the situations which are used for the purpose of literacy and mathematics training.

It is well worth looking at recent writings of UNESCO officials here. Lestage (1982) defines functionality in this way:

> A person is *functionally illiterate* who cannot engage in all those activities in which literacy is required for effective functioning of his groups and community and also for enabling him to continue to use reading, writing and calculation for his own and the community's development.
>
> Ibid., p. i

And Couvent (1979) quotes the *Declaration of Persepolis*

> Successes were achieved when literacy programmes were not restricted to learning the skills of reading, writing and arithmetic, and when they did not subordinate literacy in short term needs of growth unconnected with man. . . . The ways and means of literacy activities should be founded on the specific characteristics of the environment, personality and identity of each people.
>
> Ibid., p. 26

Accordingly it seems that we have to find methods for reaching the depths of the soul of many of our pupils in order to meet their needs for Activities and the corresponding needs for tools. Elsasser and Steiner (1977) are concerned about this. Standing on the shoulders of Vygotsky and Freire, they claim the necessity to look for methods of unveiling the child's inner speech, that is, to bring the inside into the open. The purpose of doing so is to confront inner speech by furnishing it with the improved use of written and spoken language and, I would add, mathematics.

Stubbs (1980) chooses another approach in order to achieve something of the same. He connects the relationship between the individual and her knowledge to a wider relationship, the one between knowledge and society. He is concerned in particular about the uses of written and spoken language in society. He asks questions such as:

> What is the functional justification for the existence of a written language alongside the spoken language? Why do societies and individuals often maintain two distinct modes of communication? Why do users find it appropriate to use one medium rather than another in certain settings?
>
> What are the social factors which determine how written language should be used?
>
> Ibid., p. 17

The next question, which Stubbs does not ask, is about the political justification for a distribution of the spoken and written language (and mathematics of course) which implies that group x of the population becomes literate while group y does not. Furthermore, we have to ask whether such a situation is acceptable to the educationist or not.

The thesis all this leads to is that the use of a thinking-tool in wider society should be reflected in its use within school. If writing a letter to the editor of a newspaper is part of social and political life, then such letters should be written in a school class as well, and sent to the paper. If mathematical modelling is important for analysing important matters of social and political life outside school, such use of mathematics should take place within school as well. If computers are mainly used in order to process huge sets of data in a short time such use should also be practised within school, and so on. I shall expound the above thesis further in Chapter 5.

1.2.11. *Summary*

A. I have developed a concept of *Activity* which is related to man's capacity to take care of his own life-situation, be responsible for it, and make decisions for it, together with others.

B. According to Soviet psychology of the thirties *tools* are vital for Activities, originally tools for the hand, later communicative tools and thinking tools, such as language and those of mathematics.

C. Tools are meaningful if and only if they can be related to Activity. In this case we say that they are *functional* for the individual.

D. When pupils cease to learn mathematics this will be related to their failure to relate the thinking-tools to what they can recognise as an Activity.

E. According to Vygotskian research and simple observations of children's play, we realise that they possess important thinking-tools in the form of sign-systems. These tools are functional for children's play.

F. Where Piaget tends to say "the child has not", Vygotsky tends to say "the child has, look how she uses her thinking-tools." Where the followers of Piaget, more or less in the spirit of his theories, construct tasks which direct them away from their daily Activities, Vygotsky would ask us to start there, and connect school Activities to them, having regard to the historical and social dimensions of Activity.

G. For the purposes of education, Activity theory as developed by Soviet psychology during the thirties is too limited. From the experiences reported in my Introduction, we see that the pupils are in a position where they can refuse to participate in school learning. Furthermore Freire has demonstrated clearly how silence and passiveness occur among people when they are denied access to Activities. We have to develop Activity theory so that it can include such phenomena.

H. As people of today's societies have different ideologies for their Activities, the political nature of Activity has to be analysed and explored. The necessity for this is related to D above.

I. It is implicit in the concept of Activity that we, as educators, have to include the pupils as decision-makers, planners and organisers of Activities. Moreover we have to look for methods, in particular from social anthropology, which can narrow the gap between what school by its history and current practice regards as sensible tasks for its pupils and what the pupils themselves regard as Activities.

1.3. DIMENSIONS FOR EDUCATION

1.3.1. *In Search of Educational Activity*

The ever-present problem of school as a more or less closed system within society demonstrates how dialectical reflection in education becomes indispensable. Bateson (1973) centres his grand theory of communication and learning on the simple thesis that human communication systems are *open* systems. If such a system is treated as a closed system and something goes wrong within it, it follows that difficulties can arise.

I can apply this thesis to a communication system such as a school. Consider a school which is run as a closed communication system. Say some learning difficulties emerges, and some remedial teaching is required. As the system is closed, there is a fair chance that the correction of the teaching/learning process will contain some of the same elements which originally caused the difficulties. If this is the case, new difficulties might be produced and exponential growth is on its way.

What about remedial lessons in mathematics as we might experience them? We see the "slow" learner being offered a simpler textbook, a specially trained teacher, a smaller class and so forth. But what about the location of the training (still a classroom, desks, blackboard etc), and I ask: Does not the complete setting remind the pupil of a totality which he already has experienced as nonproductive?

I shall return in detail to such problems of communication systems in §4.3. Here I note that the potentially closed nature of a school in relation to the society of which it is a proper part implies that I must consider which problems I can tackle in the context of *within* school, and which problems I have to go outside school to attack. It is mainly in the latter case that I see the necessity to politicise education. The dialectical nature of this politicisation is intrinsic to the part-whole relation of school and the society of which school is a part. In order to solve problems inside school I have to reflect on problems outside school. Conversely, problems outside school impinge on problems inside school.

Many of those who have opposed my view about the need to politicise (mathematics) education have done this from a position of defeatism, stating that a teacher is more or less committing suicide by promoting politicisation.

My objection to such dismay is to argue that most educators have a long way to go in exploiting existing possibilities to open school as a system. My argument is that school, its pedagogues, curriculum makers and so forth, have not yet sufficiently seized the opportunity to exploit the existing tolerance to politicisation of education. This should, of course, not be interpreted as meaning that educators should always stay within such limits of tolerance. The case of Freire demonstrates perfectly the necessity to challenge the lack of tolerance in — in his case — a totalitarian society.

In order to create educational tasks which are in harmony with the pupils' Activities, it might be necessary to politicise education by opening the school as a communication system. In this section I shall investigate some dimensions of Activity which can, I hope, serve as tools for the curriculum maker to create such an opening.

The cases I shall report are quite unpretentious. Experience tells us that they will occur within any liberal Western classroom. Our motive here is to clarify three important dimensions of Activity. I shall explore more ambitious projects in later chapters. The clue to such exploitation of Activity theory is the need to include biography, history and society in psychology or, as in our case, a theory of learning.

1.3.2. *The Samba School*

Seymor Papert (1980) describes the Samba schools of Rio de Janeiro as an ideal model of how he would like to see classroom activities. The Samba schools function over a year, preparing one section of the grand Carnival Parade, a twelve-hour-long procession of song, dance and street theatre. What may look like a spontaneous show is the result of laborious work, which is an important part of the life of the participants. Each group builds its own *learning environment* (in Papert's words) for the preparations:

— an idea is discussed as the foundation for the performance;
— music and songs are composed and written;
— dances are choreographed;
— costumes are made etc.

Papert stresses how different generations meet, discuss and mix socially, how the learning environment also serves as a social environment, which in its turn inspires and develops further creativity, knowledge and skills:

> Learning is not separated from reality. The Samba school has a purpose, and learning is integrated in the school for this purpose. Novice is not separated from the expert, and the experts are also learning.
>
> Papert, ibid., p. 179

Papert continues to develop these ideas as an ideal model for what he wants to achieve as a LOGO environment, or a computer culture among school children. I see at least three dimensions here which build the learning Activity, and which must be stressed to protect it from trivial interpretations.

I. *Dimension: The past and the future.* The past is present in history and traditions. The creation of the new show, with its imagination, fantasy, inventions, is only possible in the context of history, that is, in the context of what was experienced in previous years. Without having been there, we can almost hear the discussions during the planning stages, how the creations are dreamed up in terms of what happened last year, what the recent trends in the Parade have been, and what competing groups can possibly come up with this year.

What Papert does not mention is that the Carnival has deep cultural roots in the poorest parts of the population; that it is a style of living which affects the whole year, and thus has a specific cultural and historical significance for a particular group of the population. The Parade expresses more than enthusiasm; it is more than strong and colourful expression of joy and spirit; it serves also as a reading of the history of the participants for those sufficiently close to it.

To understand an Activity fully we have to know its relevant history. The same understanding is necessary for the educator when planning and giving birth to some project which she hopes will turn out to be an educational Activity for her pupils. The resulting principle is that future learning Activities have to be planned in the context of former Activities.

II. *Dimension: Narrowing and widening knowledge.* We see how a field of knowledge may be narrowed down to discussions of the properties of a thinking tool, or for obtaining necessary skills. But such a field can also be approached in a much wider way.

We can see the younger ones in the Samba school participate in the creation of new costumes. We can hear them discuss the design: the costumes are to illustrate a particular idea, and there are various interpretations. We can see them learn how to draw the patterns, to use the techniques which the particular material requires; we can see them becoming increasingly skilful at drawing, cutting, sewing and gluing various materials. All these "narrow" skills will be developed within a context which gives this knowledge a broader meaning, that is, the meaning of the Activity.

This learning of knowledge in the narrow sense, focusing on ideas and skills, will occur in a social setting where the past and the future dimensions are also present. The motivation is thus, as Freire would have expressed it, with the situation.

The Samba school is in the lucky position that it can concentrate its Activities in one project, with a unifying goal — the Parade. The skills required for reaching this goal will be obtained in the context of the goal. The interaction between the various components of the Activity, which Christiansen and Walther call for (§1.2.2.), is thus easily recognised.

The advantage in the case of the Samba school, is obviously the close connection between the various Actions (or educational tasks), and the grand project defining the Activity. When step by step we now approach the classroom which is not quite as exotic as we may think the Samba school is, we can think about the Grand Parade as a metaphor: the existence of a unifying project, where knowledge in the wide sense is present all the time, and which implies the need for knowledge in the narrow sense.

Focusing on ideas, principles, skills, i.e. thinking-tools, by analysing them, discussing them and developing them can only exist in the context of a wider project. The difficulty facing us as educators is to define this wider project, or — to understand what our pupils' projects are about.

III. *Dimension: Interpersonal and intrapersonal knowledge.* This dimension is obvious in the case of the Samba school. The final performance is the result of cooperation and individual skills, which mutually feed each other. Cooperation has not been reduced to copying or demonstrations. There may have been sharing of viewpoints, discussions, quarrels, conflicts, which in their turn promote further individual and collective learning.

We see that learning can take place in such a school without the presence of one particular teacher who is responsible for the learning Activities. There will be communication from old to young, from experienced to inexperienced, from skilled to unskilled in a variety of settings.

An analogue which comes to mind is the two different serving systems which are used in restaurants. In European restaurants the waiter usually has some tables which are his sole responsibility. In most Chinese and Indian restaurants each waiter has responsibility for each table: no-one will pass any table without taking action if necessary. The Samba school is an exotic example for the ignorant Scandinavian educator. It is probably less exciting, more a part of daily life for those participating in it. Still it shows some of the dimensions which we have to take seriously when analysing Activity theory from the perspective of education. We now take a step nearer the mathematics classroom.

1.3.3. *The Laboratory*

Hoskyns (1977) shows us how much can be achieved by quite small means in an ordinary school context. True enough, these "small means" are some of the most unfair things with which we can confront teachers. Hoskyns works outside school-hours: he opens his physics laboratory for his pupils then. One of the ideas behind so doing is that he can create Activities which can provide positive feed-back to the learning of the core curriculum, i.e. knowledge in the narrow sense. To mention just a few of the Activities in the laboratory:

- fitting a hi-fi amplifier and speakers into the lab with a radio and a turntable, leading up to establishing a record loan club;
- building a discotheque for use in the school hall and local clubs;
- encouraging all those with an interest in electronics to build and test circuits, getting guitar amplifiers and cabinet speakers and let groups of students use a room to practise with microphones and tape recorders.

As Hoskyns himself expresses it:

> I am suggesting that in so far as our students find their physics both stimulating and rewarding, literacy and numeracy become integrally involved in the student's activities, rather than problems for yet new committees of inquiry.
>
> Hoskyns, ibid., p. 164

There are still a few other qualities to be mentioned here. The laboratory was obviously a valuable resource for the pupils outside school hours. It mirrored something of the total life situation for the pupils, something about their housing conditions, possibilities for indoor leisure activities, which were important activities in their environment and so forth. Furthermore this emphasises what is one the basic themes of this book, the rapidly changing environment of children all over the world, and the implications of this for education.

1.3.4. *An IOWO Project*

IOWO stands for Instituut voor de Ontwikkeling van het Wiskunde

Onderwijs. In its lifetime, it was established in 1971 and was laid down in 1982, it served as a commando center for mathematics education in Northern parts of Europe. In *Five Years IOWO* (1976), a script dedicated to its director Hans Freudenthal, the first thing we read is a sign: *Attention Children*!

Now, eight years later, we are in a position where we can take a critical review of what we studied at that time. My example will be the project called *Building a Bungalow*. It centres around the cube. The children investigate reticulations of a cube; they stock four cubes in different ways, counting faces; they draw them in two and three dimensions, and so on.

They build cube bungalows on a recreation area. They discuss which plot would be the most preferable according to a variety of factors. Here indeed are mathematical activities, mathematical ideas, mathematical learning. Everything takes place in a realistic setting: there are bungalows, plots, building costs, negotiations etc. And still I refer to it as activities in mathematics, not using capital "A".

I argue that the theory of IOWO does not convince one that it really considers the dialectic dimensions between knowledge and the learners as required by Activity theory.

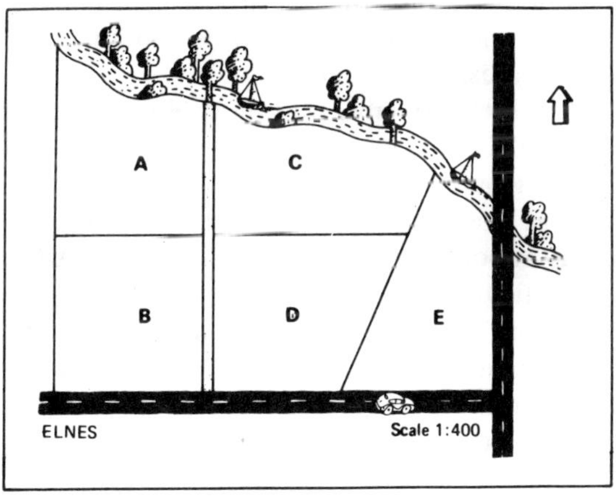

Fig. 1.3.1.

The project is, as always with IOWO, the uniting of some powerful mathematics and children's work (discovering, experiencing, discussing) in a context which the group *anticipates* will be interesting for the pupils.

But how do we know about this interest? How can we know that such a project strikes the pupils in their historical and social context as I have discussed it so far? Well, we can argue that children usually build when they have the opportunity, and often we can see them build houses as well. And children live in houses, and they know that some day they will have their own place to live. So is not this formulating the future in terms of the past?

My answer to this is, yes — perhaps. Of course there will exist pupils who live happily in bungalows, and who think it is nice to reproduce these bungalows in order to investigate mathematical ideas connected with them. Of course, several of these pupils will take it for granted that they some day will be owners of their own bungalow, so that the project makes sense like that as well.

But the very same children may be angry on the day they have to manipulate the blocks, because they have been refused permission to play football on the lawn between the houses, or on the road nearby. So where then could they play their game?

Or they might come from high blocks of flats which are composed not of four cubes, but of 400, and they hear their parents complaining about the situation, not knowing what to do avoid it, and it is almost a kilometer to the nearest playground, and . . .

This IOWO project certainly fulfils most of the criteria which can be set for a project which it is intended will turn out to be an Activity for the pupils engaging in it. The only way in which it might miss the target seems to be this: how can we be sure that the kind of knowledge focused upon is important for the pupils involved? And *important* in the context of Activity also reads "important in the historical and social context of the pupils". It is exactly this point on which our interest will focus more and more closely in what follows: by which theories and methods can we ensure that mathematical knowledge becomes part of the knowledge which the pupils themselves recognise as important? The first step will be to take a further look at the three dimensions described so far.

The project illustrated on the following pages is a slight, but important variation of the bungalow project of IOWO.

The children (Class 1 of a Norwegian school), built their own environment. This does not necessarily guarantee that a learning Activity emerges. But the possibility for this increases as they build their *own* environment. In this case, the children raised such discussions as: Where can we play? Where would we have liked to play? Can any compromise be made with the adults? Where is the most dangerous crossing? Can anything be done (Figure 1.3.6.)? The presence of an Activity will, to a certain degree, be signalled by the contribution of the children's own discussion to the learning situation.

Further: let us plan a children's village. Children here contribute from their own experience and knowledge, as an important part of the project. The role of the teacher in such an Activity is complex, as she has to organise and communicate knowledge without directing too much.

1.3.5. *The Past and the Future Dimension*

David Hansen had been very lucky indeed. In the very same year as he graduated from college, he got this job in the school in a quiet village outside his native town. The first year David was to have the 9-year-olds. He scanned through the maths textbooks, and found that some simple geometry and statistics were part of the syllabus. Watching the road outside the school windows, David decided to do a traffic project.

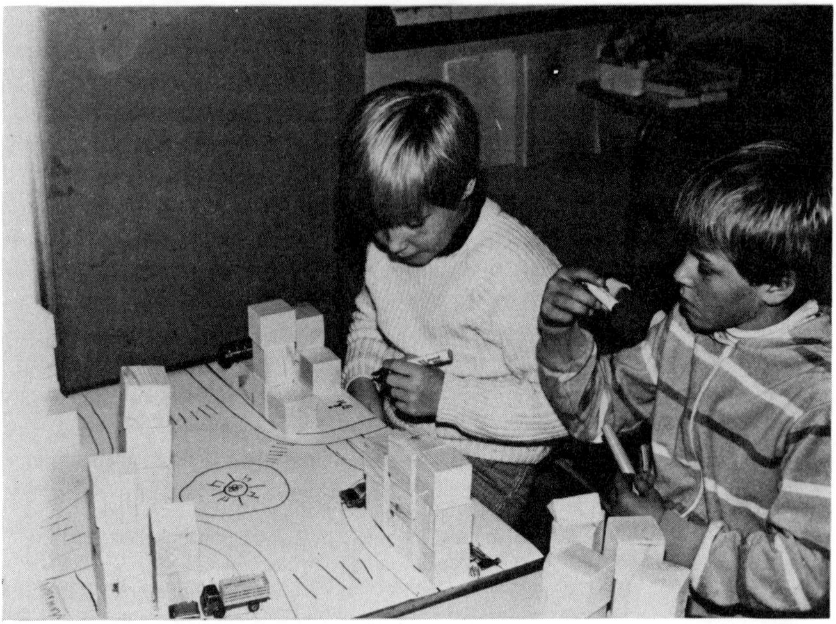

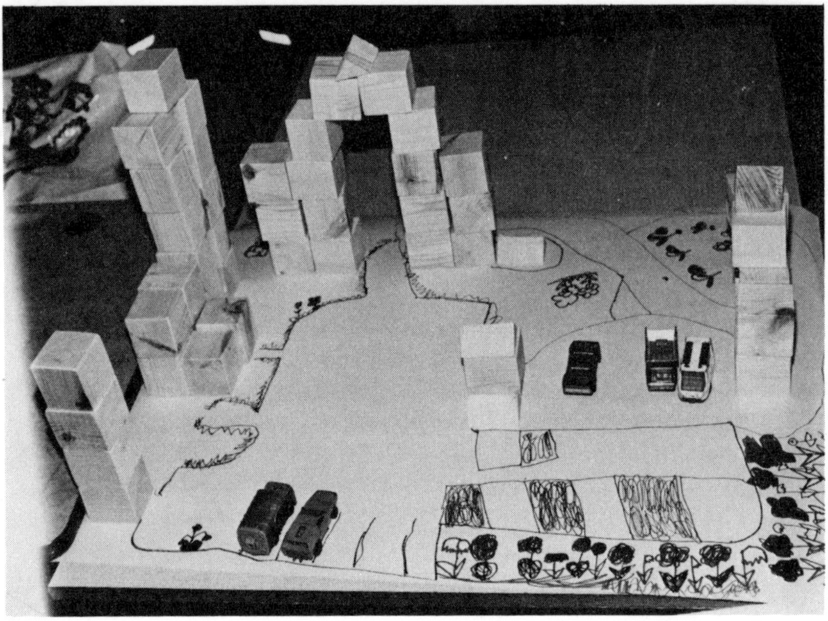

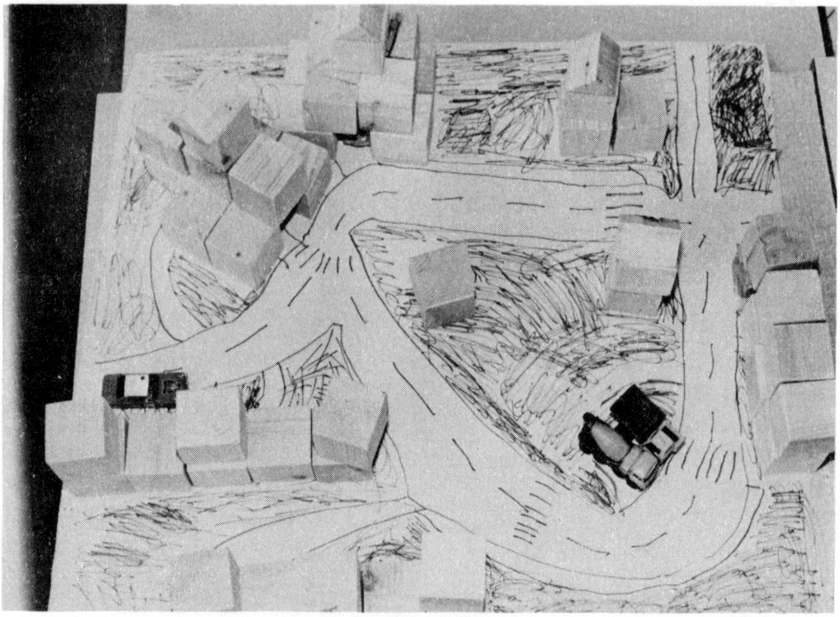

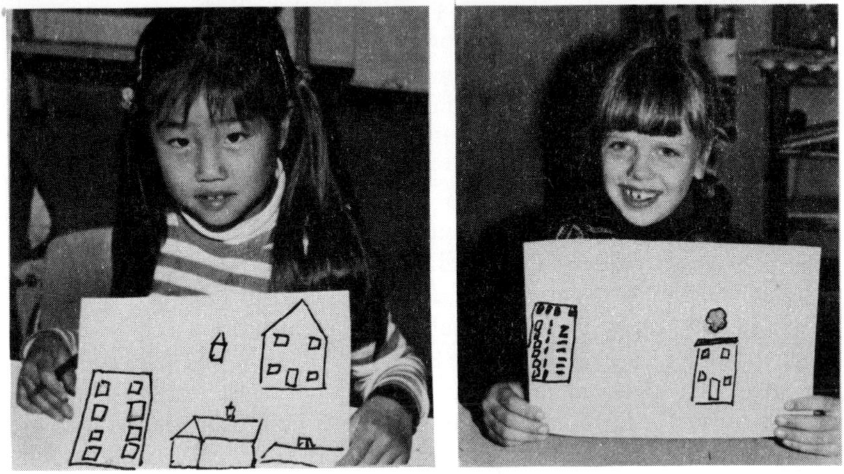

Figs. 1.3.2.—1.3.10

The class spent lots of time drawing road junctions and designing crossings for pedestrians. Some of the pupils positioned traffic lights as well, and at David's suggestion they constructed an algorithm for the timing of the light during the peak period and the rush hour. That made some of the pupils suggest that they could go out and collect some statistics on queuing at the nearby junction to see if the timing there could be improved.

The project lasted for 2 weeks and David was quite happy with everything, until Sue came up. She showed him a petition and asked him to sign it.

"What is this about, Sue?"

"Oh, you see Mr. Hansen. After Market Street became restricted to one-way traffic, there has been so much non-local traffic on our road. Even the lorries use it now. And it is not built for lorries, you see."

"I didn't know that. This is Valley Drive, isn't? Several of you live there."

"Oh yes Mr. Hansen. Kate, Eva and me. And some of the boys, of course. We used to play there before, but we are not allowed to any more. Dad says the road was once built for horses, and that cars should never have used it anyway."

"That is really bad, Sue."

"Yes it is, Mr. Hansen. But me, Kate and the rest of us are counting the lorries. Mummy and Mrs. Jackson count them in the morning. And we are to do some statistics and send them to the authorities so they can find another road for the lorries. Would you please sign, Mr. Hansen?"

Invented story for the purpose of our theory? Yes. Rare story? No.

It is my experience that teachers who develop a sensitive attitude towards the historical and social setting in which they perform their profession regularly come across such cases which prove an excellent foundation for educational tasks promoting pupils' Activities.

Mr. Hansen almost had the golden egg in his hands here. The problem was that he did not know what to reach out and grasp. And above all — no part of his teacher-training had attended to this.

The impact of the individual's history on his Activities is related to various fields of his life and to various levels. In the case of David Hansen we can see the role of previous incidents on present Activity: there exists a road with a certain traffic history which conflicts with today's lorries, and this conflict causes concern to those living on the road. There exists, however, another historical level, which is equally important for learning. That is the level of strategies for Activities.

Any individual, family, group and society inherit certain strategies from former generations about how to cope with reality. In the case of the trunk road it is not in the hands of every social group to plan and carry out statistical investigations. Nor is it the tradition of every social group to undertake militant action by — say — blocking the road.

At the level of language I have already mentioned the impact of socio-linguistics (§1.1.4.) on communication: different social groups employ different linguistic registers for the storage and communication of experience. Bernstein's findings relate to the experiences of the mathematics teacher:

"Why are we proving this, Teach? Why can't we just use the result? Why bother with all this proving — we can see it works all right, can't we?"

When working class families assimilate middle-class values, when Third World countries seek Western ways of living, they have to rely on strategies for doing so which are rooted in tradition and history geared to former ways of living. It is this contradiction which Marx so brilliantly describes as the dialectics of history: human beings create history, but they do not create it according to their free wishes; they do not do it under their own circumstances; they do it under circumstances transmitted to them from the past. The novice who has learned a new language will first translate this language into his mother tongue. He masters the spirit of the new language the moment he automatically, without reflection, feels at home in the new language and thus forgets his mother tongue.

Marx's conception of language includes both the ideological and political aspect, with political action as the necessary and unavoidable goal. The implication of the above for the mathematics educator is that it is hard to see how educational tasks in mathematics can be designed without consideration for the learners' inherited strategies and traditions related to the storing and communication of knowledge.

The worry of the pedagogues of our time in general is the severe problems facing large groups of young people as they try to cope with a rapidly changing world. What if the changes come too quickly for the required development of strategies for survival?

It is an unfair exaggeration to say that schools never use the history of their pupils as the basis for Activities. It is practised quite a lot. The children will participate in projects concerning religious events, excursions to important place in the near environment, there will be plays, bazaars and so on and the parents will be invited to participate. Several headteachers and teachers have a policy of making the school a cultural force in its social environment. There are, however, two questions to be raised here:

(i) To what extent does such a project genuinely orient itself towards the history of the pupils?

(ii) To what extent does the project function independently of the rest of schoolwork, such as language and mathematics learning? That is, to what extent are the most important thinking-tools exploited in such projects?

To take (i) first. In the field of multi-racial education (Stone 1981, BBC 1981), lots of things have been done as educators have recognized the importance of illuminating the history of immigrant children in order to combat racism. On the other hand, we experience ambitious headmasters supporting the organisation of string-quartets in areas where folk music is alive or the children themselves develop new traditions of music (usually various forms for rock music these days).

The Swede Sven Lindquist (1979) has written a book which demonstrates the point: *Dig where you stand.* The title says everything. The simple advantage of the principle stated in the title is that as all people have a history, and as all people live in some sort of social context, we shall always strike something when we start to dig. The problem is just to get started.

Experience tells us that the task is not as difficult as it may appear. It is not mainly a question about serious and laborious research of the social-anthropology type. Teachers who mix with non-teachers socially, teachers who meet the parents of their pupils in the supermarket, sports clubs, social clubs, are in a most favourable position to learn about their pupils. In short: teachers who face the parents in off-duty situations can learn significant information of use in their teaching. The same is the case of course with contact of that kind with pupils.

Naturally this has to go along with an observant attitude and an interest in other people's lives. And we see the failure of most teacher education to consider such simple principles, orienting the students towards learning theories where the social and historical constituents of human life are usually neglected.

1.3.6. *The Narrowing-widening Dimension*

The study of the properties of a thinking-tool is its study in the narrow sense. In mathematics we study the structural properties of thinking-tools, their domains of definition, and their possible generalisations. The exploitation of their potential for application is the study of them in the wide sense.

It is the educators's treatment of the dialectics between the study of thinking-tools in their narrow and wide sense which is significant for whether the learner will perceive the thinking-tool in the context of her Activities or not.

In §1.2.2. I referred to Christiansen and Walther's analysis of the relationship between educational tasks and Activity. They build some of their work on the thinking of Davydov and Markova (1982—83). These two Soviet pedagogues analyse a concept of educational activity for schoolchildren. They structure such activity into three components.

1. The basic unit is the child's understanding of an educational task, which appears as the study of a thinking-tool or knowledge in the narrow sense here. Davydov and Markova describe such a task as being closely related to an interesting (theoretical) generalisation, bringing schoolchildren to the mastery of new methods of action. Educational task situations, being basic units of Activity, require the child's acceptance of the content matter for its own sake, as something worth studying there and then.

2. The performance of educational acts. This corresponds to the Actions comprising an Activity, as described in §1.2.2. Here the thinking-tools as developed by educational tasks are used in a wider sense.

3. Finally, there is the pupils' own performance of acts of control and evaluation. Remembering that an Activity is the pupils' project, connected to an object and a motive, the educational tasks and the educational acts will be evaluated accordingly.

I do not see the conceptual structure of Davydov and Markova as sufficient. They describe educational tasks as basic units, and they relate them to larger projects, called educational acts, which ultimately serve the educational activity.

But I find little or nothing about criteria which help me as an educator to initiate projects (that involve thinking-tools) which will be recognised as Activities by my pupils. This relates to the pupils' rationale for learning, as I shall discuss it in §4.1. Regarding the historical dimension, which Davydov and Markova miss, I see the concept of a "tool" as developed historically by Vygotsky as indispensable. Tools used in their narrow sense are dependent on their use in the wide sense (that is in Activities), and the latter is dependent on the historical and social factors, as also stressed by Vygotsky and Leont'ev.

The important thing for the Western educator to learn from Davydov and Markova is their emphasis on the interrelationships between the various components of an Activity, not only in its planning stage, but also as something which emerges for the pupils as they work with the project. The success of the project will depend on whether the pupils discover the thinking-tool in the context of its use in the wide sense or not.

Papert (op. cit.) is concerned about creating computer environments, seeing programming as a thinking-tool. He stresses a very important goal here, which is often neglected by learning theorists: the child's use of the thinking-tool implies the child's reflection on her own thinking, that is in the words of Bateson: the child's metalearning about the tool. In the case of computers, by simple programming children teach the computer how to think, and thus embark on an exploration about how they themselves think:

> I began to see how children who had learned to program computers could use very concrete computer models to think about thinking and to learn about learning, and in doing so, enhance their powers as psychologists and as epistemologists.
>
> Papert, ibid., p. 23

But here problems start to arise. It is difficult to see how creating a computer environment by itself would be sufficient to encourage learning — and learning about learning. We can see (as we have seen) young children programming their home-computers, constructing mile-long programs for war-games accompanied by changes of colours and music, using repetitions and subroutines, with the same sort of enthusiasm which we can observe when they exploit games in amusement halls. It seems here that Papert thinks it is sufficient merely to bring in the appropriate technology, with the status of being futuristic, and leave it at that. If this is the case, we are back to the "process"-oriented education, where the content of the situation is of no interest and it is the quality of the process which count. Papert's neglect of the historical dimension becomes quite clear when he writes: "the intellectual environments offered to children by today's cultures are poor in opportunities to bring their thinking about thinking into the open, to learn to talk about it and to test their ideas by externalising them." (Ibid.)

I shall not disregard the fact that this statement contains some truth. The major experience however, is the contrary, especially for those social workers who have the streets and street-corners as their major working place. The experience of street-children's reflecting on their own situation and on their own reflection is unanimous among these social workers. And such reflection is necessary, because for many of them it is often a question about to be or not to be. There is, of course, a stage in the career of many of the street-children where they cease to be, as in the case of heavy drug addicts.

So bringing in any thinking-tool, however fashionable it may be, cannot be sufficient in itself: something more has to be added to curriculum planning, something which includes the pupils as historical and social objects. It is in such a context that the thinking-tool can be functional and make pupils realise its importance in its narrow sense, for further investigations, developments and generalisations.

Introducing the narrowing-widening dimension of knowledge brings in another advantage for the curriculum planner. In order to grasp the point, I have to introduce Bateson's concept of metalearning (the concept will be analysed in §4.3.). A series of learning situations will imply metalearning about the characteristics of these situations, which equips the individual with some learning strategies for the next situation. Metalearning is primarily long-range learning, developing and transforming over time.

The point is that if the pupil experiences the narrowing—widening dialectic function, she will gain some experience of the power of the thinking-tool. If so, we can expect her to take on board another tool, only experiencing it in its narrow sense. The principle is a risky one, as it may be tempting to overdo it. If the pupil did take one, perhaps she will take the next one as well, and the next and the next, and sooner or later we discover that she said "no" some weeks ago. The metaphor "thinking-tool" is a good one. If we think of carpenters' tools, most of us will see the use of a hammer, a saw and an axe, and we appreciate having them in the house. When someone one day gives us a carpenting tool x_1 which we have never seen, and cannot see the use for at the moment, we will probably appreciate the gift anyway, knowing that we may sooner or later need such a tool, and perhaps x_1 as well may come to use. Who knows? But as our kind donor constantly provides us with new tools, x_2, x_3, . . . , x_c, although their use is explained to us, filling up our shelves, we will probably some day ask what is really going on.

In a certain way we are here back to dimension I, about the significance of the past for the future, as we discuss the pupil's learning history for his future learning. One conclusion from this is the need to take care of the metalevel in communication with the pupils. This implies repeated discussions about knowledge, following the various dimensions, in parallel with the pupils' construction of this knowledge.

Davydov and Markova make a call for such discussions when they stress their concern about the interrelations between the various components of an Activity. Similarly, it is what I call for when I stress the necessity to move dialectically between the past and the future, and between knowledge in the wide and the narrow sense.

1.3.7. *The Inter-intrapersonal Dimension*

This dimension is clearly the most difficult to handle for curriculum planning. I have previously stressed that Vygotsky built his theory in the young Soviet state, where no ideas about conflicting goals among its people could reign. Nor is ability a concept in Activity theory. Activity theory as it is developed by Vygotsky and Leont'ev leads us to suppose that the individual pupil will have to exercise her skills in the context of the collective (group) in such a way that the interpersonal Activity will nourish all sorts of abilities in the collective, for the sake of the common goal. In other words: in an Activity every pupil would contribute her efforts and grow by so doing for the benefit of the collective (group) and its goal.

Now the Soviet State has obviously changed its course and the nourishment of individual mathematical ability has become a goal in itself (Krutetski 1976).

The problem most educators will face today is that the pupils will have different goals for their education. This implies that what becomes an Activity for one pupil will not be that for another. So when the educator invites participation in an Activity, the pupils will not necessarily participate. One aspect of this is that the society can be a competitive society where cooperation in Activities can only function up to a certain point: the pupils will sooner or later catch sight of the implications of the coming examination. Cooperation for competition is a paradox which will easily will lead to a double-bind for those involved (see §4.3.).

Furthermore we shall face the difficulty that if, for a while, we can disregard external examinations (after all they do not dominate every school hour) our pupils might have different ideologies. That is, they might have different motives for participating in an Activity. This problem is somewhat easier to cope with, and its solution will be found by bringing such differences into the open, rather than by disregarding them — by *politicising education.*

This situation is not all ideal. We do not like to think about little ones as people who can actually have different and even conflicting interests. And we like to practise cooperation and harmony, and not bring the nasties of the outside world into classrooms. But somehow we have to cope with conflicts when they are real. As educators most of us face situations which include dilemmas such as (competition versus cooperation) and (ideology x versus ideology y). We have to face rather then disregard such problems, as they demonstrate the limitations of our curriculum planning.

In Chapter 5 I hope to demonstrate how opposing ideas, viewpoints, ideologies, as discussed and challenged by the pupils, can bring in some freshness, energy and motivation for the further learning of mathematics. So far I have merely located problems rather than pointing to their solutions.

CHAPTER 2

MATHEMATICS AS A LANGUAGE

Think first, then code!
Copper and Clancy (1982).

2.1. THEORY

2.1.1. *Introducing the Topic*

According to the theory outlined in the previous chapter *language* is the basic thinking-tool of the human being. In this chapter I shall pursue this paradigm in order to examine its role for the thinking-tools of mathematics.

The analysis will be exploratory in the context of Activity theory. Recent research about the role of language in mathematics education has given new insight to the mathematics educator, an insight which in many ways provides a dramatic opening for new strategies for mathematics instruction. The empirical findings of this research disclose that our pupils are in possession of much more mathematical knowledge than the educators have been aware of. The problem is that they store this knowledge and communicate it in other coding systems than the standard systems authorised by the curriculum.

Thus the concept of *translation* of a coding system, i.e. a language, from the one in the learner's possession to the standard one, will be considered in this chapter.

The conception of the learner as one who possesses much more mathematical knowledge than we have foreseen opens up new conceptions of the *dialectics of the classroom* and the teacher's direction of learning. If it turns out that the pupil really has mathematical knowledge related to the topic of a particular lesson, we have to consider what kind of strategies this can generate for the teacher.

This conception of the learner as "one who knows" also implies that we have to sharpen our analysis of *non-verbal* mathematics as contrasted with *verbal* or *formalised* mathematics. The paradigm about language as basic for reflection invites a conclusion which reduces mathematics to language. Such a conclusion is, as I shall argue in the next chapter, obviously false. To examine the precise role of language for cognition in general is one of the chief enterprises of cognitive psychology, and is beyond the scope of this book.

I am further painfully aware that my analysis lacks the precision of *the language of mathematics.*

Is it the coding systems that are characteristic of mathematics? The equations, formulae, matrices, tables and so forth? Or should we include the structures of the spoken language which embody the codifications, that, is the *registers* of mathematics which Halliday (1980) describes?

My solution to this analytic deficiency is to ask the reader to examine my use of the *language of mathematics*, in order to assist the location of this conception. In the forthcoming exploratory analysis of the role of language in mathematics I shall merely focus on the communicative aspects of language in learner/teacher relations, mainly considering learning situations where different coding systems are present. This is relevant to Bauersfeld's concern about the distinction between the "matter learned", the "matter meant" and the "matter taught" (§1.2.2.), and the standpoint that these three features of an educational situation should, if possible, in some way be in harmony.

2.1.2. *Preparation for Written Language*

We start with nursery education. Parallel with the development of speech as a sign system, children develop other systems which function on the basis of speech. Vygotsky draws attention to *gestures*:

> The gesture is the initial visual sign that contains the child's future writing as an acorn contains a future oak. Gestures, it has correctly been said, are writing in the air, and written signs frequently are simply gestures that have been fixed.
>
> Vygotsky, 1979, p. 107

Again we have a parallel between phylogeny and ontogeny: the development of number-systems and their signs on the basis of finger-counting (and other systems of body-counting), and their pictorial representations as written signs, were similar to the shape of the gestures:

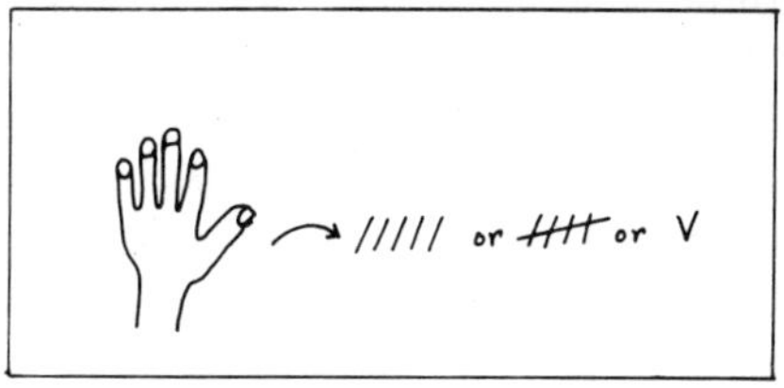

Fig. 2.1.1

until the dominance of the Hindu-Arabian system.

We also see the child actively use his hand when counting: "one"-"two"-"three"-.

We notice the transition to written signs: Five fingers are represented by /////, or five counting bricks are represented by 00000.

The point here is not only the understanding of a particular sign system, such as "how many are there?", but the understanding of the functionality of the system; that is, a system which is in the child's possession for use according to his needs. This implies the understanding that important information can be stored and communicated by the use of written signs, and even more important, the child's understanding that he himself posesses the potential to do so.

It is the discovery of the significance of the number symbol as a written sign, the recognition made by the child that he is in a position to use it for his own goals, which we shall look for.

Children's drawings become important in this connection. Disregarding the aesthetic features of drawing, we can observe that they contain information as a written sign. If the pupils' minds are occupied with yesterday's birthday party and they are making a drawing of it, in their drawings most of them will communicate the age of the birthday child (number of candles on the cake), and the number of guests (the various presents will usually be drawn).

When drawing his family or another important group of people, the child will also create a written sign showing "how many", a sign which can be used for storing and communicating. By collecting such signs, interchanging them, children can compare, add, subtract, and sort: how many 2-families, 3-families and so on.

The educational trademark here is not so much the connection of an incident with a written sign, as the child's discovery that he possesses written signs, that written signs are good for something, that he can use them, thus learning about the *functionality of a written sign-system*.

The psycholinguist Hermione Sinclair makes the same point when she summarises recent research on young children's acquisition of language:

> Nobody seems to think that children already know how to add, subtract, multiply and divide before they come to school, and that all they have to learn is to do pencil-and paper sums. On the contrary — in most countries arithmetic is taught as if the conceptualizations of written arithmetic operations were the same as their written symbolization.
>
> Schools do not seem to envisage that the conceptualization of addition, subtraction etc. may be a cognitive task separate from that of writing equations, and that the latter may present difficulties of its own.
>
> Sinclair, 1983, p. 9

According to this we should not be as quick as we often are to explain the mathematical failures of pupils in terms of *conceptual* deficit. Often the observed failures can be the result of language problems, i.e. problems of using the expected standard language. I shall provide some case studies about this in §2.1.7. which will serve as examples.

2.1.3. *The Ogden-Richards Triangle (O.-R.)*

In 1923 the book by Ogden and Richards *The Meaning of Meaning* was published. They set out to investigate the relationships between language and meaning. More generally, their concern was the relationships between a sign system and meaning. Thus they focused on the same problem as Vygotsky, although from a philosophical point of view, and not a psychological one like Vygotsky's.

They built their theory on a simple model which was later much praised and much criticised for its simplicity. The model can be represented as a triangle, showing the relationships between an incident of the real world, its mediation and its expressive sign:

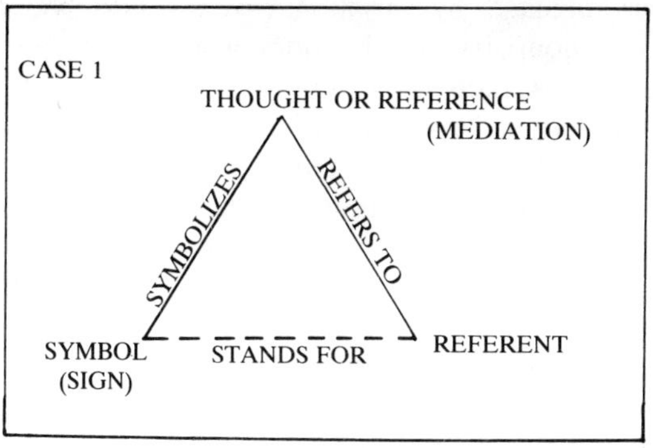

Fig. 2.1.2.

We see that the base of the triangle has been dotted, demonstrating that every use of a sign is the result of some mediation by the individual. As Vygotsky also would have stressed, the sign has an objective, general appearance: its interpretation, "thought of reference" in the words of O.-R., is subjective. There are a few immediate implications of this theoretical construct which are of importance for all kinds of language education, including important fields of mathematics education:

(i) Two individuals may interpret signs differently as they meet them in an educational (and other) context.

(ii) Different languages, or sign systems, will organise meaning differently.

In (ii) we touch on sociolinguistics and the Sapir-Whorf hypothesis, and we see ahead a chain of high mountains.

O.-R., contemporaries of Saussure (and the young Piaget and Vygotsky), strongly reject Saussure's conceptions of the sign and the signifier. They refer to Saussure's use of *language* as a ready-made construction, predetermining speech.

O.-R. reject such a position, regarding language as a structure emerging from the use of signs by the individual in his social context. Language is not some total set of linguistic elements in which the individual develops his speech repertoire. Language is rather the growing repertoire of the individual as he continually constructs new uses of signs.

According to this view the child should be as free as possible to build his own language, by variation, inventiveness and fantasy. This process is to be nourished in situations where it is important for the child to express his motives. It is the possibility of such linguistic creativity and functionality which provides much of the reason for the popularity of the O.-R. paradigm among modern language educators.[1]

The parallel with the language of mathematics may seem dubious. But we shall soon face problems concerning the various forms of representing knowledge employed by pupils and the traditional expectations of mathematics education about efficiency and standardisation. To what extent is mathematics education a question of letting the pupil build his own mathematical language; and to what extent is it a question of guiding him into a ready-made field of language use?

This dilemma is of course not an "either-or" dilemma. It is a "both-and". And we shall look for a proper balance.

There are, of course, weaknesses with the O.-R. model, as seen from our position also. Malinowski (1923) already points to the major problem in his afterword to their book, when he stresses the necessity of relating language use to context.

Malinowski, a social anthropologist, studied native cultures in terms of their language behaviour. This explains his enthusiasm for the O.-R. model. Furthermore Malinowski felt that he could not interpret such behaviour without knowing in what kind of situation the language was used. This point has become crucial in modern sociolinguistics as well, use of language being studied in terms of *registers*, registers being related to context (Edwards, 1976; Halliday, 1980).

This weakness of the O.-R. model leads us straight to the next criticism of O.-R.: that they analyse the relationships between sign and meaning in terms of static situations, and not in terms of Activities. Activities are of course context-bounded. Consider a set of "bananas". In terms of the reasoning of O.-R., we would have

— the existence of the set;
— the meaning given to it;
— the various signs representing it.

Obviously the meaning will be subjective.

A worker who works for Chiquita in Latin America will probably symbolise a set of bananas in a different way from that of children in Scandinavia. Still the situation is a static one. Bananas exist, and so what? What are we going to do with them? Count them, add them, eat them? Produce them, import them, budget them, buy them, fry them?

In what sort of Activity do the bananas really appear?

This demonstrates that O.-R. worked in the context of philosophy and not in education. To them there just existed referents, their mediation and their signs.

2.1.4. *The Høines Triangle*

The teacher faces much more complex situations than those the O.-R. triangle can describe. The teacher will face situations where the learner decides not to participate in the learning situation. He will face situations where the learner judges the learning situation differently from what he himself does. And so forth. The teacher meets such situations as:

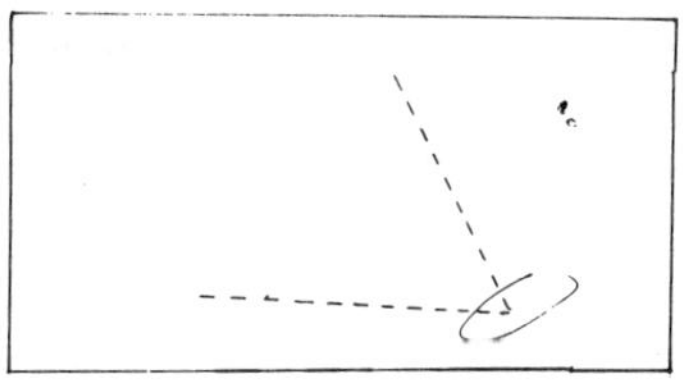

Fig. 2.1.3. The pupil does not engage.

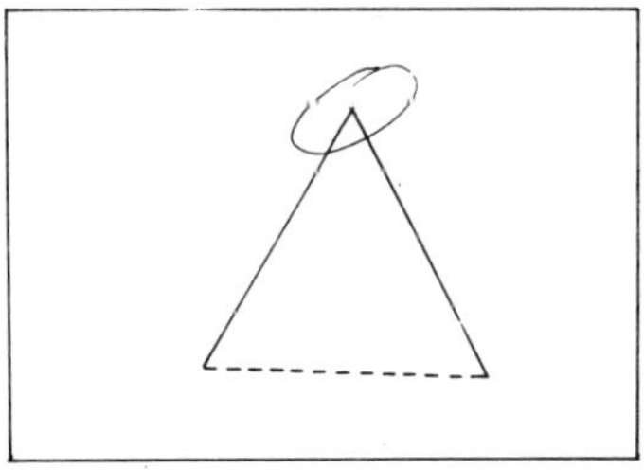

Fig. 2.1.4. "I write it, he says I must."

I shall still keep the O.-R.-triangle as it provides a tool for examining the use of language in a classroom situation. I have at least two possibilities for extending its use such that it can function as a tool for Activity theory.

The first possibility is to conceptualise the triangle as an Activity-triangle, modelling not only one particular incident of language use, but rather the total language use in an Activity. This suffers from the fact that in an Activity there will be various uses of languages and of different languages (such as mathematical language and the spoken language). Thus most of the point of the O.-R.-triangle is lost.

I would therefore rather use the O.-R.-triangle as an *incident* of an Activity, thinking about an Activity as a set of language uses, where each can be modelled by the triangle.

From here on I shall use the expression '*Høines-triangle*' or '*H.-triangle*,' as Høines (1983) uses the O.-R.-triangle in this way.

The H.-triangle will now thus serve as a snapshot of an Activity. The triangle will now be an interpreting tool for the teacher of the Activity of the pupil. It will function as a tool for observing "the matter learned" in relation to the "matter taught" and the "matter meant". By using it the educator can get some idea about the course of the pupil's Activity.

A single H.-triangle will not always be sufficient to tell us which Activity the pupils engage in. By using a series of triangles, one can some closer. It is like interpreting the content of a film by choosing an ordered string of pictures: we can obtain a fairly good idea of what it is about, but we will not grasp its artistic flavour.

The strategies for choosing such strings are worth a discussion in their own right, especially in the context of teacher education. It is in this way that teachers usually observe their pupils' learning Activities. It is especially so when they associate the pupils' work with knowledge in the narrow sense, i.e. when they concentrate on conceptual analysis and the development of skills. Bishop and Whitfield (1972) discuss the teacher's decision-making in situations where the teacher faces an occurrence of an Activity, and provide an illustrative list of teacher's problem-solving in this field.

2.1.5. *Variations of H.-triangles*

In the following cases of H.-triangles, it is taken for granted that the pupil engages in an Activity which complies with the educational tasks presented to him by the teacher, and that the H.-triangles picture an occurrence of this. From now on I shall use the term "language" about a sign system in general.

Case 1
This is the basic case, as modelled in §2.1.3. This is a necessary, but not sufficient condition, for a case of language use by the individual which is *functional.* The keys for sufficient conditions are to be found in the relationships between language and Activity as discussed in §1.2.

Case 2
Two or more functional languages are present.

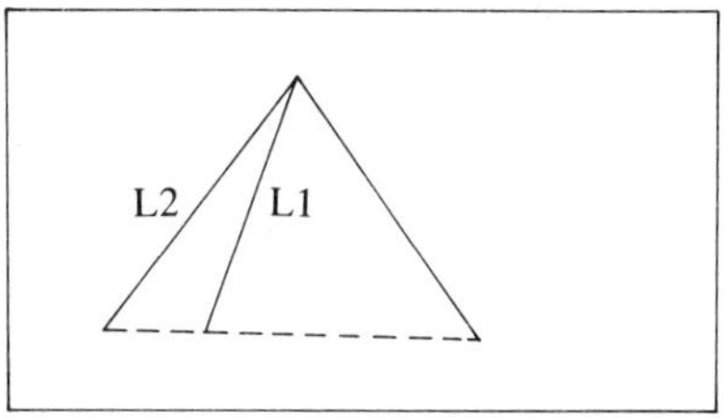

Fig. 2.1.5.

Here L1 and L2 are two languages, which function in *parallel* for the individual. This implies that there is no need for the individual to translate from one language to the other in order to obtain meaning. The present author is an example of this situation. English, being my second language, was originally translated into Norwegian, in order to provide meaning. As English becomes more familiar, particularly as a result of staying in England for a while, English is not translated any more: the inner language will be English, and thus English can function in parallel with the first language.

The situation is also characterised by the shift from the use of a Norwegian-English dictionary, to the sole use of Roget's Thesaurus, which implies that when in need of linguistic expressions, the search for help takes place *within* the new language.

Another example of the same is in the case of *programming.* When programming in a structured language such as Fortran or Pascal, one is recommended to write the algorithm in daily, commonsense language. The criterion for such writing is that it has to be communicable: that is, by reading the algorithm other people should be able to perform its operations. This algorithmic language is then translated into the programming language.

Some will recommend that a *pseudocode* be employed as an intermediate language (Prather 1982, Cooper and Clancy, 1982). As the student of programming gradually becomes familiar with the programming process, he can immediately write his programs in the programming language without being dependent on the use of everyday language. In this case the programming language will function in parallel with the everyday language.

It is interesting for the mathematics educator here to notice how the theorists of programming delay the coding process to the very last minute: first thinking, then coding. Can the reason for this be that programming is a brand new science vital for the modern society, a society which cannot afford instructional blind alleys in this field?

This process of familiarisation with the new language obviously includes the *discovery* of its language properties, as well as the individual's *decision* which will be dependent upon comfort, motivation and familiarity with the new language.

To approach the classroom, we can illustrate Case 2 with the examples of finger mathematics, which is prevalent among 6—7 year-olds, and written mathematics. The children can do simple arithmetical operations by counting their fingers. In parallel, they can do the operations by writing the corresponding mathematical signs. One goal for education at this stage is thus to develop parallelism between the two languages, as a condition for later independence of the finger language.

We remain with the problem that two parallel languages will generate non-isomorphic meanings. Taking the point of MacKenzie that people of different nationalities who employ different languages cannot "by any means bring themselves to think quite alike, at least on subjects that involve any depth of sentiment; they have not the verbal means" (O.-R., 1923, p. 1), we see that case 2 is something more than reducing the use of a language L1 to a new, parallel language, L2. This relates also to the Sapir-Whorf hypothesis, to be discussed in §3.2. To be precise, we ought to have drawn another top vertex of the H.-triangle, in the neighbourhood of the original one. This is obvious in the example about the programming language. Such languages will usually be built up by means of blocks, each block getting a label, so that the main program will bind the blocks together by calling them up by their labels. No-one would structure their daily language in such a way when describing the solution to a problem. The mediation, as Vygotsky would have put it, would take place according to other patterns when using everyday language, than when using programming language directly.

But again we must have in mind that we are only highlighting moments in the history of an Activity, not the Activity itself. What will prove important for teaching, and what I shall analyse further in the next case, is the analysis, investigation and discussion of the properties of the sign-system under consideration.

2.1.6. *Case* 3

The substitution of one language for another by *translation.*

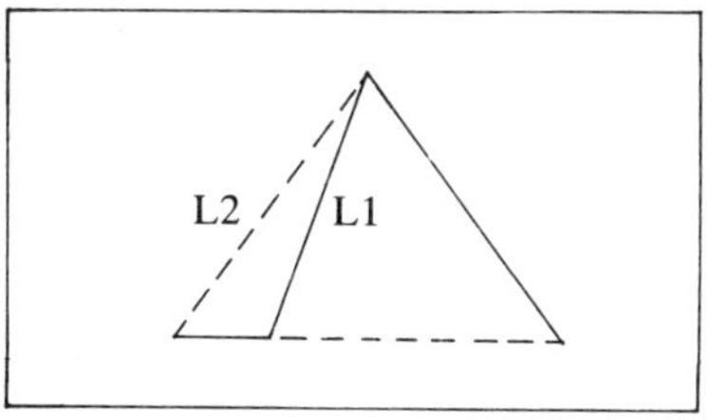

Fig. 2.1.6.

Here L2 is not yet functional, and has to be translated through L1. L2 is a substitution of L1: L2 goes through L1.

This is what most language teaching is about, and we are now analysing mathematics learning in the same way. It is what we are doing when we teach children their first use of numerals, when we teach them how to model a written problem by means of mathematical signs, when we teach them how to use standard algorithms on the basis of their own algorithms and so forth.

When children are to learn mathematics by *writing* mathematics, they will possess knowledge of how to count (speech), add up small sums, of relations such as 'greater than', 'less than', 'equal to' etc. If they do not yet have such knowledge, in terms of spoken language, then it is sound and familiar pedagogy to provide them with such language. It will then be time for translation into the new written language by means of the functional spoken mathematical language for the purpose of obtaining a Case 2 situation.

In the case of problem-solving, the problem will usually be given in a written form, by means of words, and has to be translated into mathematical sign models. The translation process is here much more difficult to perceive, and also is lost in many classrooms. That is, the teacher "forgets" that at some time a Case 3 situation will be present, and teaches according to a Case 2 situation by writing the algorithms straight onto the blackboard by means of mathematical symbolism.

I shall return to the discussion about the use of standard algorithms in §2.3. Here, just to give the reader a hint, a typical example is the slow learner who can write $20 + 20 + 20 + 3 + 3 + 3 = 60 + 9$ rather than

$$\begin{array}{l} 23 \times 3 \\ \;\;60 \\ \;9 \\ \;\;69 \end{array}$$

or some other standardisation.

2.1.7. *Two Case Studies*

Case of Robert (*aged 7*)
Robert does his sums:

$$\begin{array}{l} 1+2= \\ 3+1= \\ 2+2= \end{array}$$

He looks at the paper, his mind apparently far away. Is it lack of concentration, or does he have poor concepts? Robert knows that one krone plus two kroner is one less than three plus one kroner. He even does not need his fingers to know that the first sum gives him three and the second four. The concepts are alright. The problem is rather that Robert is not getting sufficient help in translating from his functional spoken mathematics language to the written mathematics. The teacher takes his translation capacity for granted: he is under various pressures, and he takes the situation to be Case 2 when it is Case 3.

Robert does need help in translating. His apparent lack of concentration may be secondary to this. It is time to become familiar with the new language which may be his basic requirement.

Case of Elin (aged 7)
A written test in her first year at school:

Draw as many strokes as the numeral shows:
3

Elin writes: ///

Write the numeral showing how many:
///

Elin does not write "3".

Again the teacher may think that this is a case of poor concepts. But a further examination of Elin's language skills might reveal that there is nothing wrong with her concepts. She might have a lot of experience of dealing with "3": she knows, for example, that it is one more than two; she has experienced that two moves, three each, correspond to a six; she knows that it is the exact fare on the bus. And she can employ primitive signs for numbers, she can demonstrate "3" by drawing three strokes, but the standard sign system for numbers is still not familiar to her.

So she becomes worried when facing the second problem in the test. And her teacher may have learnt to interpret this situation as an incident of poor number concepts rather than as an incident of a language which is not yet functional, coexisting with a more primitive written sign system which is functional at this level of number operations. If this is the case, we have Case 3, which should not be treated as a Case 2 situation of the H.-triangle.

2.1.8. *The Relationships between L1 and L2*

It is an implication of Vygotsky's theory that in order to develop a new functional language to obtain a Case 2 situation, the original language (the base) has to be functional. In most cases the original language will be speech. This relates to the discussions of §1.2.7.

The following analysis will rest on this assumption. This implies that if speech is not functional, it is no use attempting to develop new languages. Failure to observe these principles frequently occurs in classrooms, often as a result of the constraint of lack of time. In sum, the error being made is to treat a Case 3 as a Case 2 situation.

The research into the mathematical ability of bilingual children and mathematical ability is of particular interest here. This research is mainly connected with immigrant children, whose native language will be L1, and the language of the new country L2. The latter will also be the language of instruction. We have cases of several Third World countries where English will be the language of instruction at quite early stages in school.

Austin and Howson (1979) report such research, and have to conclude that bilingualism does not necessarily lead to poorer results in mathematics. From their report, however, it is difficult to tell the quality of the L2 of the individuals as compared with the L1. Was L2 functional, and was L1? Had L2 to go through L1? And if the latter is the case, is L1 geared towards the kind of relationships which are mathematised in the L2-society? The latter is difficult to examine, and I shall postpone discussion of it to Chapter 3.

Dawe (1983) provides some powerful research about the strengths of a developed L1 for mathematical reasoning within L2. He tests hypotheses derived from the theory of Cummins, which asserts that a cognitively and academically beneficial form of bilingualism can only be achieved on adequately developed L1 skills.[2] On the basis of his research Dawe suggests that:

> whatever the medium of instruction it would seem likely that mathematical reasoning in the deductive sense is closely related to the ability to use language as a tool for thought. In the case of bilingual children this involves competence in *both* languages.
>
> It has been clearly shown that the ability of the child to make effective use of the cognitive functions of his first language is a good predictor of his ability to reason deductively in English as a second language. It would seem likely that this result may also be true of a wider range of language pairs than examined in this study. Furthermore, the sociolinguistic distance between the two languages may be an important factor to take into account.
>
> Dawe, ibid., p. 349

We notice here the concept of *sociolinguistic distance* between two languages. We shall have to consider it for a wider understanding about the relationships between an L1 and an L2. We will meet it repeatedly from now on.

Dawe (ibid.) uses a test battery as his research tool. With this he tests competence in L1 and L2. Furthermore he also has a test for deductive reasoning in mathematics in English. And he makes a great leap forward with his work. But studies on the impact of the relationship between L1 and L2 are still missing. By testing L1 skills independently of L2 skills, to some extent we can only guess whether a Case 2 or a

Case 3 situation is present and about the implications of such cases for the learning of mathematics in an L2 language. It seems that close observation of individual pupils could be a good way to proceed in order to obtain such knowledge.

A challenging thesis for teachers of mathematics will be that it is better for a pupil to practise a functional language which is improper (according to formal standards) and inefficient, than not to practise a language at all. This principle is realised by an increasing number of language teachers, who follow the guidelines of dialect projects. They start off by teaching the children to read and write in their own dialect in order to move into the formal language, rather than rush into the formal language filling the children's books with red marks. Still quite a lot of mathematics teachers will hesitate anxiously when they face an algorithm by a 14 year old written as

$$\begin{array}{ll} 48 \div 4 & \\ 1 & 4 \\ 1 & 4 \\ 5 & 20 \\ 5 & 20 \\ \hline 12 & \end{array} \quad \text{rather than} \quad 4\overline{)48} \atop 12 \quad \text{or} \quad \begin{array}{l} 48 \div 4 = 12 \\ \underline{4} \\ 8 \\ \underline{8} \end{array}$$

Remembering the discussion about Case 2, that the development of a new functional language is as much a question about being familiar with the new language and *deciding* to use it, we can hardly persuade or force a pupil to employ it.

2.1.9. *Case 4*

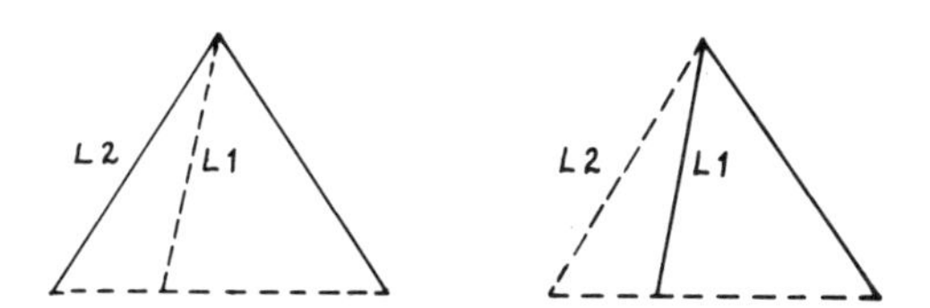

Pupil's choice of language   Teacher's expectation

Fig. 2.1.7.

In this case the language use is functional, but a different language is used by the pupil from the one the teacher expects. This is not to be confused with the case where the pupils expressed themselves in a different Activity from the one their teacher expected. Recent research demonstrates that Case 4 is much more common than we have recognised so far.

Booth (1981) suggests, on the basis of her findings, that there are two systems of mathematics coexisting in the secondary classroom: the formal taught system, and a system of child methods which are based upon counting, adding on, or building up approaches, and by which children attempt to solve mathematical problems within a "human sense" framework.

Hart et al. (1981) report on a large-scale project on children's understanding of mathematics. They conclude that the children for the most part did not use teacher-taught algorithms. Some rules would appear, however, as part of the children's repertoire. They further stress how children to a great extent adapt the algorithms they are taught, and replace them by their own methods. It was only when their own methods failed that they saw the need for employing the rule.

The most convincing research in this field has been by the American educator Robert B. Davis.[3] His innovative project, The Madison Project, was set up to liberate children's potential for employing their own forms of mathematical language. One is struck by the richness and variety of these forms when an appropriate setting is provided. Ginsburg (1977) provides similar data, and hints at the phenomenon of children's secret arithmetic, the one they more or less consciously keep secret from their teachers as it is not of the "standard" type. Sometimes these methods follow familiar and expected routes, sometimes they reveal structures which produce systematic errors. Sometimes taken together they build a complete algebra, producing both correct and incorrect results. Ginsburg (1982, 1983) develops the *clinical interview* as a method for spotting such algebras. Such interviews are performed by letting the pupil think aloud as he solves a problem or performs an algorithm. The interviewer interjects questions when approriate for determination of the child's method.

Those teachers who practise this method in quiet moments in their classrooms usually obtain important feedback for their teaching.

2.2. BEGINNING MATHEMATICS

2.2.1. *Starting Up*

It was in 1978 that Marit Høines and I decided to forget all about Piaget for a while and embark on an innovation concerning Beginning Mathematics. At that time she was going to start afresh with a new class of 6—7 year-olds — the starting age in Norway.

She worked in a well-established school in the countryside outside Bergen, and the school's reputation was such that no parent would expect that anything hazardous could take place there. Later the parents of the class, with one exception, all supported her approach. The exception was a mother who practised as a teacher at the same level in another school. As everything turned out, we know from experience that the approach probably rescued two or three pupils from the hands of the special educator.

During the first year of our innovative work, we had not realised the depth of Vygotsky's use of the term "activity". We soon became impressed by the children's capacity to express their mathematical experiences in a written form once we had relieved them of the usual formal expectations. This made us concentrate on their expressiveness through written symbols, rather than on the context in which they were used. The demonstration by the pupils of their capacity to express experiences, is a crucial evidence of the presence of an Activity.

I well remember the first project. I brought four youngsters to the roadside. For 5 minutes they were (a) to count the number of cars driving towards the town; (b) count those coming from town; and (c) write down their countings.

They were organized two by two. In each pair one was to do the counting, reporting to the other who was to write it all down. There were first some moments of hesitation. Write? But Vygotsky had said that children at this age could store information in written language, and if they could do it in Soviet Russia, they should be able to do it in Norway as well. So I shrugged my shoulders to their question about "how to".

Then there was a member of the group who got the idea, usually writing

 or 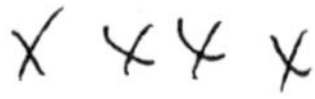

for each car passing by.

Little by little we experienced signs like this:

Their interpretation is left to the reader.

The next great moment came when we returned to the class. Before the next group of four was allowed to rush out to count cars, the results had to be written on the blackboard for interpretation, counting and discussion. The signs were collected in groups of five, and there was a poster to keep the records. The variation in the figures was discussed: when were there most cars — in the mornings, in the afternoons? A ceremony in the nearby church, was that spotted? And what about the nearby car-ferry?

After some weeks Marit started a session by asking how many ordinary people a car could seat. The answers to this were not the expected five, and this tells something about children's counting culture. Their mode number was 7. The explanation is clear for everyone who has observed Mam bring and collect the small ones from the kindergarten, or has done his share to and from a birthday party or the sports hall.

Anyway, we did some negotiation, and arrived at the wanted five. Marit then grouped the children in fives, and each group took a place near the walls in order to do some driving. The children sat two in the front and three in the rear, and suddenly there was plenty of space in the room. So the 24 in the class were transformed to four full cars and one with an empty seat, and what could the situation be on the road, and what could be the explanation of that situation?

Then the children suddenly found themselves counting cars with just one person in, transforming their statistics into a statistic of fives.

The following graphs became very impressive, the shrink from a graph showing 30 cars with one person, to a graph showing 6 cars with 5 people in. And what about the buses which passed all day and which could seat — how many? — remember to count next time you travel, and report back. The children at this stage still had not got at their disposal other written signs than the ones they themselves had invented. They were in the second month of their very first year at school.

Fig. 2.2.1.

Fig. 2.2.2.

2.2.2. *On Method*

Some of the pupils were familiar with the Hindu-Arabic symbols when they arrived at school. This, of course, caused a problem of differentiation for their teacher. We encouraged them to use their own inventions of symbols; and, as it turned out, they had no problem in doing so. The pupils would compare and discuss their symbolic inventions. When they questioned us, we gave the pupils the answer that it was up to them which symbols they would employ.

All the time their teacher stressed the use of the spoken, mathematical language. The experience in these classes, is that there are still some who are not familiar with the spoken "number language." So the pupils were all the time provoked to count, compare and conclude. The school day is full of opportunities:

— How many are absent? And yesterday? The day before yesterday?
— What is the development? Can there be a flu epidemic around?
— Is it mostly children from one specific area who are absent?

. . .

— Are your books really full?
— How many shall I get?
— What are you saying, Kate, have you only two pages left?
— Shall I get new books for all of you who have less than 3 pages left?
— Will they be finished tomorrow? How do you know?

. . .

— Oh dear. Not that again. Fluoride treatment. All right. You will have to go four by four to nurse. Eight minutes for each group. Will we make it before lunch? How many minutes will each group have then?

. . .

— Kenneth, could you come up and show us your method? I see. So you drew the clock did you. And counted like that. Very clever. Does anybody else have a method he would like to show?

If we call the standard written mathematical language L2, we can see this two-language user as L1: the spoken mathematical language, and the written one which built on the children's own inventions. If we call the last two respectively L1 and L1′ we obtain the following picture:

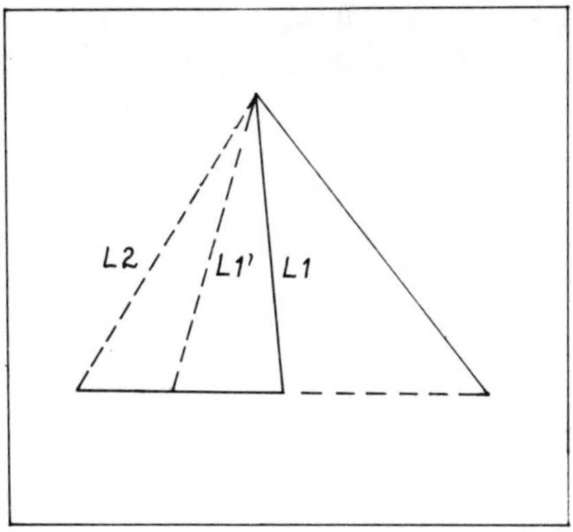

Fig. 2.2.3.

spoken language (L1) conveys own written system (L1′)

own written system (L1′) conveys normed system (L2).

It is an implication of almost all that has been said about Activities, that a fixed timetable is not the best framework. We therefore did not follow one. During the project work, there was a focus both on the use of written and spoken mother-tongue, as well as on mathematical language. There was a stress on environmental studies. As the observant reader may note, the projects — as environmental projects — offered possibilities for integration with other school subjects as well.

The mathematics textbooks were used for treating knowledge in the narrow sense: the various exercises were done when there were empty spaces in the daily schedule. It was part of the weekly planning to provide suitable space for such exercises.

2.2.3. *About Projects*

A project-oriented approach like ours demands more mathematical knowledge of the teacher than usual. Besides having knowledge about mathematical symbols and their use, she needs knowledge about the possibilities mathematics offers for the pupils at the various levels she teaches.

According to our experiences from the in-service training in Bergen, many teachers have quite firm prejudices about what the pupils at a specific level are and are not capable of doing. Those who carried out some innovative projects have all been surprised at the capacity of their pupils to "take" the thinking-tools which the project required. Another problem which occurs when pursuing the use of mathematics as a set of tools for Activities, is that when mathematics educators have explored the possibilities of various thinking-tools for classroom projects, the selection of the tools usually has been in the context of mathematics, and not in the context of possible Activities for the pupils.

This is the familiar problem about whether new mathematics should be learned through projects, or projects should restrict the use of mathematics to thinking-tools already learned. In the first case, situations will occur where the pupils face some problem and where they need a tool from mathematics. This problem, however, will occur on the pupil's path and not on the educator's. Thus the teacher must be prepared to offer some tools from her own collection. It is not assumed by doing this that *all* the thinking-tools shall be presented to the pupils in this way. The problem is to obtain some good projects which demonstrate the importance of thinking-tools from mathematics.

The teacher is here provided with a difficult task, which he might not be prepared for. We can expect him to take advantage of the pupil's obvious needs and interests by initiating a project.

But I ask here for something more. I ask the teacher to exploit the possibilities of mathematics in order to let the pupils obtain new information about the project matter. Sometimes the teacher, by his metaknowledge of mathematics, will have to explore his thinking-tools together with his pupils. The task is not as demanding as it seems. Actually, the various journals for mathematics education, written by teachers for teachers, are packed with examples of projects, where the teacher has invited his pupils to do some adventurous thinking into mathematics. This shows that where confidence and motivation are present, capacity follows. So a huge potential is apparently present.

2.2.4. *Some Projects of Beginning Mathematics*

In the following pages I will report some projects. I will only hint at the use of mathematics. The projects were carried out in various classes, in the first term of the year the pupils started schools.

Project: Counting dinners
Sunny August. Children's second week at school, just getting acquainted with the routines.

"Miss — we went fishing yesterday. We got hundreds."

"Fantastic. Anybody else go fishing yesterday?"

Twenty stories about recent fishing and thousands of fishes were offered, before Miss had to proceed with *her* doings. But soon after:

Fig. 2.2.4.

"Could you make a drawing of when you went fishing?"

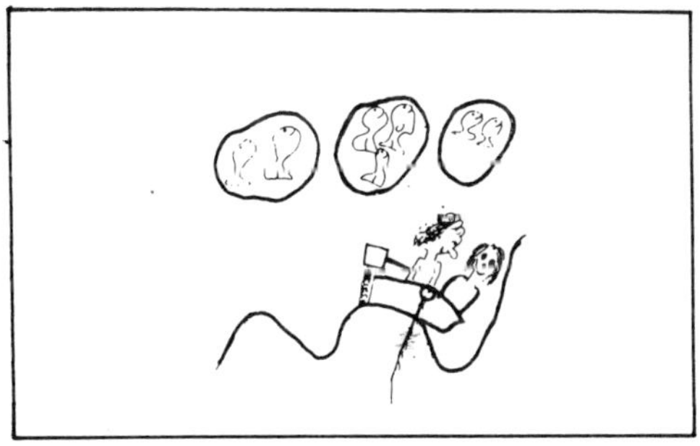

Fig. 2.2.5.

"First we got those two, then three and then another two. We got seven altogether. Two and three and two are as much as seven."

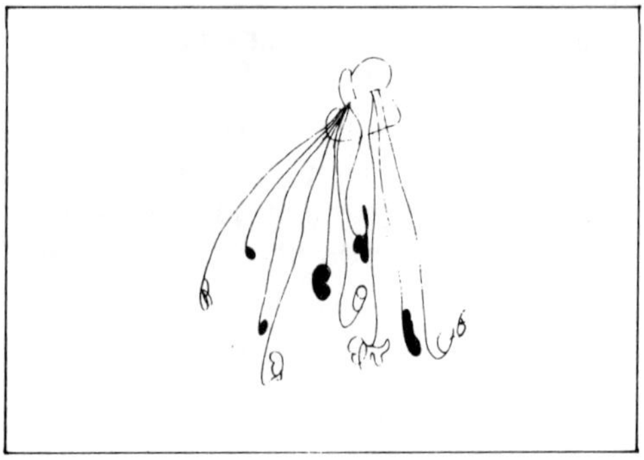

Fig. 2.2.6.

"Here are five and here are five. That is ten."

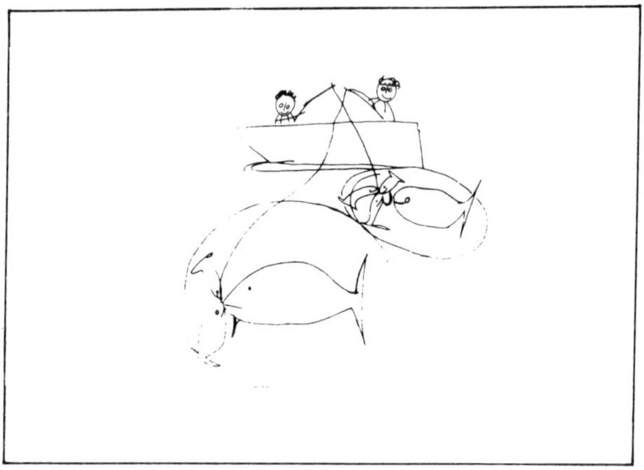

Fig. 2.2.7.

"I got three. My sister four. She had one more than me, but that big cod, *that* was mine. Oh — we had many dinners from that one. Fifteen kilos it was."

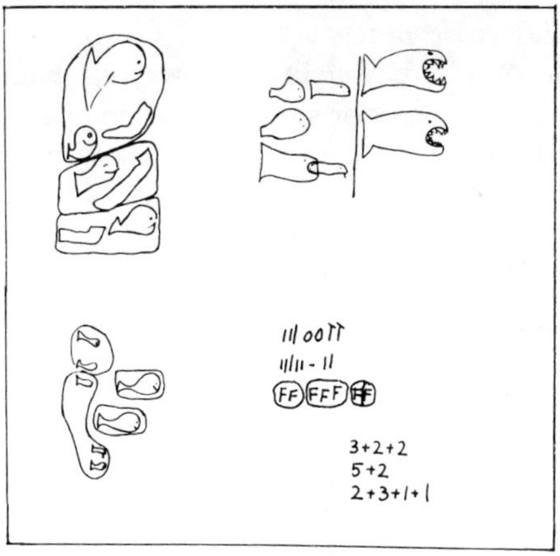

Fig. 2.2.8.

Children write their sums

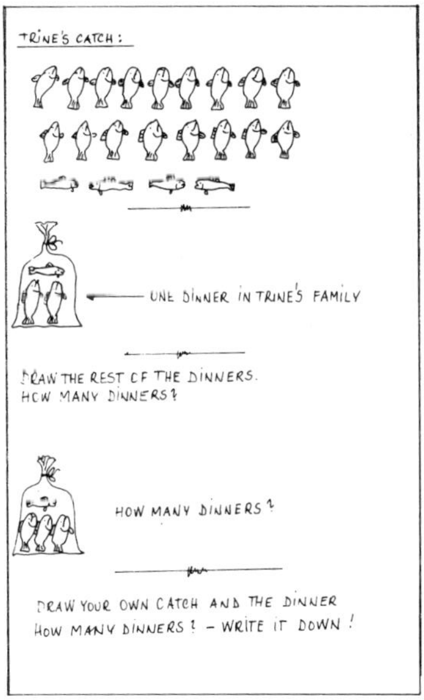

Fig. 2.2.9.

Teacher's systematisation

Project: The nearest environment
Make a drawing of where you live. Draw your family as well. How many? Make a sign on the door showing how many. Yes, of course you can make a sign showing how many adults and how many children.

How many are there in Alice's and Erik's family altogether? Write it down.

Tomorrow it is Saturday. Will you go picking berries? The wood is full of blueberries. What — do you hate picking berries? Not you Ben, nor you Martin?

Now listen carefully: I want you to make a drawing of your house again. This time you will draw those at home in the window of your house, and the rest I want to see going for berries. Could you vary this situation?

This latter problem was intended for the clever ones. It turned out that everyone found all the partitions into two integers of an integer up to ten, so we had to carry on:

Sometimes not all want to go together. Some want to go fast, some want to go alone, some want to go far. Can you find all the possibilities? Make drawings.

Fig. 2.2.10.

Tasks: *Draw those living in this house*:

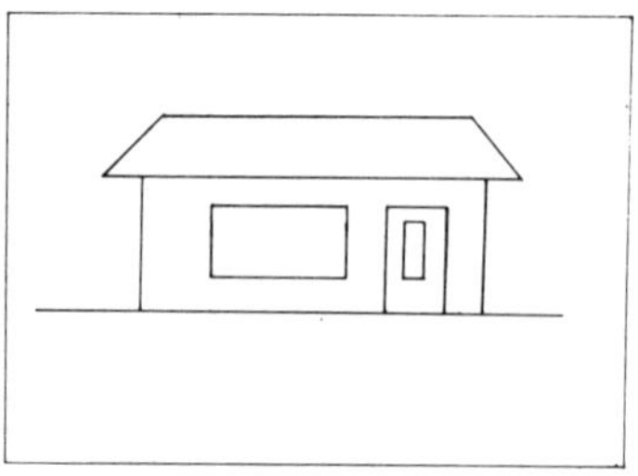

Fig. 2.2.11.

Today some people are out for a Sunday walk. Those at home are seen in the window. Draw those who are out.

Fig. 2.2.12.

Fig. 2.2.13.

Monday—Tuesday— . . . —Sunday. Who is home?

Project: Parents' occupation
Starting point: Can you make a drawing of Mum and Dad at work? What mathematics must they know? Is there any work where we do not need mathematics?

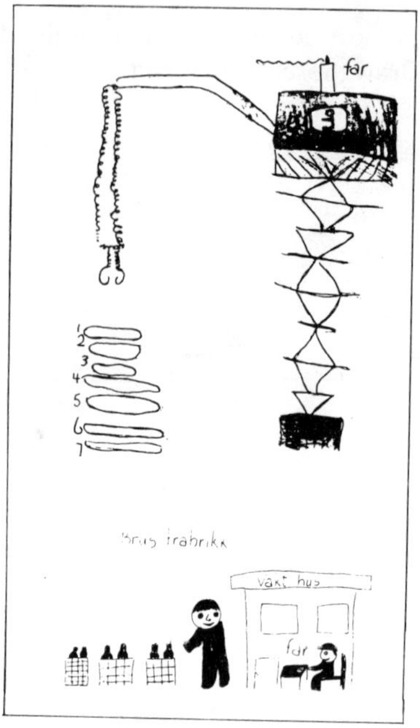

Fig. 2.2.14.

Fig. 2.2.15.

2.2.5. *Summary*

So far much has been plain. I have analysed language use in mathematics at the elementary level. The pupils have done some counting, some graphs, some classifications and groupings, some studies of variation in data as a function of time. There has been a specific stress on the function of mathematical symbols which were developed for these purposes.

So at least we have some assumptions that, by such an approach, our pupils have obtained some metaknowledge about the use of mathematical symbols, some concepts about themselves as people who are in possession of the capacity to use such symbols, and the possibility to use their own symbolic systems, if the standardised signs are forgotten or not understood.

I have made the claim that in order to develop such metaconcepts, it is necessary for the pupil to discover the functionality of the symbols, which in its turn implies that the pupil has to recognise the tasks which his teacher involves him in as a component of an Activity.

My next step will be to analyse algorithms in the same way. Algorithms in the general sense are for more than doing simple arithmetic. Algorithms, as we will study them, are the way to perform tasks in mathematics which require a set of operations rather than a single one. Obviously the treatment of language will play an important role here as well. The pupils' metaconcept of an algorithmic language will tell him something about his own capabilities and possibilities: is he an individual who is in posession of a language which is a tool to solve problems or is he restricted to the use of the standardised algorithms and thus solely dependant on those?

2.3. ALGORITHMS

2.3.1. *Definition*

I shall consider an algorithm in the usual way as a finite set of actions which, performed in a specific order, solves a problem. For the time being I am not interested in algorithms as designs for computer programming or in the use of hand calculators. The interest is in the child's construction of algorithms, his use of them, and following this, a growing metaconcept about algorithms.

I shall, though, adopt some of the principles emphasised in information theory:

— the algorithms ought to be efficient in use; and
— when presented as an algorithm, the communicative element should be considered.

The latter requirement implies the criterion that when someone reads an algorithm, he should be able to solve a similar problem by following it. The definition of an algorithm used here is clearly inspired by information theory (Cooper and Clancy, op. cit., and Prather, op. cit.).

But as we shall soon see, this definition fits particularly well with recent research on "children's methods", as opposed to "standard methods" in arithmetic. The definition above will give the various children's methods which can be accorded status as algorithms. We can build our theory of instruction accordingly: rather than neglecting children's methods, we shall have to take them seriously and discuss them; and we want to develop them and translate them.

2.3.2. *Algorithmic Actions*

Relating Activity theory, in particular Vygotsky's theory of thought and language, to the previous analysis of *algorithms*, we obtain the following corollaries:

A. As an algorithm consists of a set of actions, there can exist several algorithms for solving a problem. Finding out about 5 twelves, can be done by

$12 + 12 + 12 + 12 + 12$
$10 + 10 + 10 + 10 + 10 + 2 + 2 + 2 + 2 + 2$
$24 + 24 + 12$
$10 \times 12 \quad 120 \quad 120/2$

B. Just as we distinguished between actions and their symbolic representations, there exist several ways to represent one and the same algorithm. Returning to the 5 twelves, we can write:

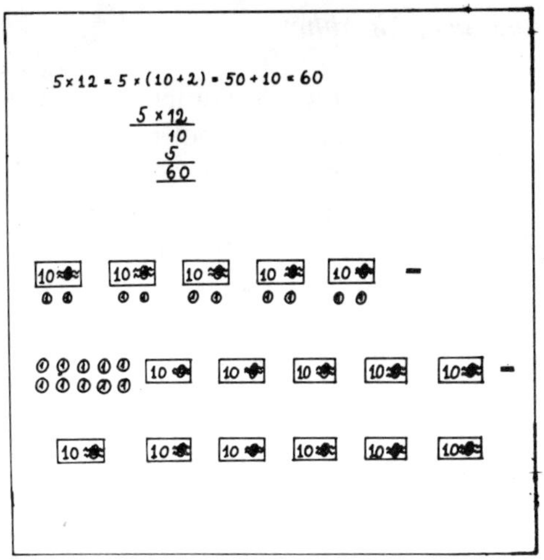

Fig. 2.3.1.

as symbolic representations of the same algorithmic actions. The considerations made here tell us that we must distinguish between *algorithmic actions* and the *algorithmic language* employed to represent these actions.

In order to know something about children's potential for solving a problem and thus constructing an algorithm, we have to know something about what languages they have at hand, i.e. what their functional languages are. *Language* is here used in the general way, as a sign system, including a language of manual actions.

I refer here to the principles stated in §2.1. on how to develop a new functional language, on the basis of one which necessarily has to be functional. The following examples may serve as an illustration:

> Three kids, aged 15. They never managed the division algorithm. In the remedial class their teacher makes strenuous renewed efforts but in vain. But did you ever see them share a packet of cigarettes (20), or a box of 200? Or 200 kroner? They will cope.
>
> And thus they do have an algorithm for division. They will communicate this algorithm when sharing. What is their language, and how can this be translated and generalised? We see here how we also have to take the enactive mode of representation, or representation by manual work, into consideration for algorithmic learning.

2.3.3. *A Metaconcept of Algorithms*

It seems quite paradoxical that we do not take advantage of developing a metaconcept of *algorithm* among school pupils. As teachers we can put much effort into problem-solving activities, analysing various algorithms for a particular problem. It is somewhat more doubtful whether we have traditionally utilised the opportunities thus offered to discuss algorithms as such with the pupils. This implies the analysis of a particular algorithm not only in the context of the present problem, but also in the context of *algorithm* in general.

The paradox occurs as the pupils have obviously developed some metaconcept anyway. An increasing amount of research demonstrates that pupils are aware that they employ a specific *method* when they solve a problem.

I have previously referred to Hart et al. and Booth on this (§2.1.6.). Brown and Burton (1978) emphasise the same thing, by pointing out that pupils' errors in elementary arithmetic are not random mistakes that occur because the pupils have not learned the proper method. In a large number of cases they were able to predict which incorrect answers the pupils would obtain as they applied their incorrect procedures in a consistent way.

Schoenfield, quoted by Brown and Burton, stresses how pupils are active agents in the construction of their own knowledge. Borrowing a conception from programming he refers to the "bugs" which can be the by-products of the pupils' attempts to perceive regularities in the world around them. Our experience is that when children get the opportunity, they demonstrate creativity, imagination and variation when they produce their own methods. These methods can be based on endless, laborious repetitions, or they can be elegant and operative solutions to the problems they face.

Davis (1967, 1980), Goffree (1982, 1983), and ter Hege (1983) show this in cases where children produce algorithms for common problem-situations, such as those requiring long subtractions, multiplication and so forth.

Research such as that done by Erlwanger (1973), Ginsburg (1973) and several others reports intriguing case studies of children building complex arithmetics of their own, thus demonstrating some of the conflicts caused by the contradiction between the inventiveness of the children and the control factors working against it. Probably the most powerful research approach towards problem-solving during the last decade is that of information processing, based on the exploitation of theories of artificial intelligence (Newell and Simon 1972; Ginsburg 1983 and Gregg 1977). Davis (1983) builds his paradigm for a theory of mathematical cognition focused on *observation* and *description* of mathematical behaviour, usually by the use of detailed taskbased interviews. The observations are related to a *postulated* theory, which describes the pupil's information processing metaphorically. The metaphors are drawn from the theory of artificial intelligence: "frames", "procedures", "subprocedures", and so forth.

Such approaches can easily be thought to be based on the idea that the "human brain is to be interpreted as a computer". On the contrary, by borrowing the metaphors from computer language, one achieves research tools which operate with the child's premises: one maintains a position from which a deep analysis of the pupil's mathematical thinking becomes possible (Davis op. cit., Davis and McKnight 1979).

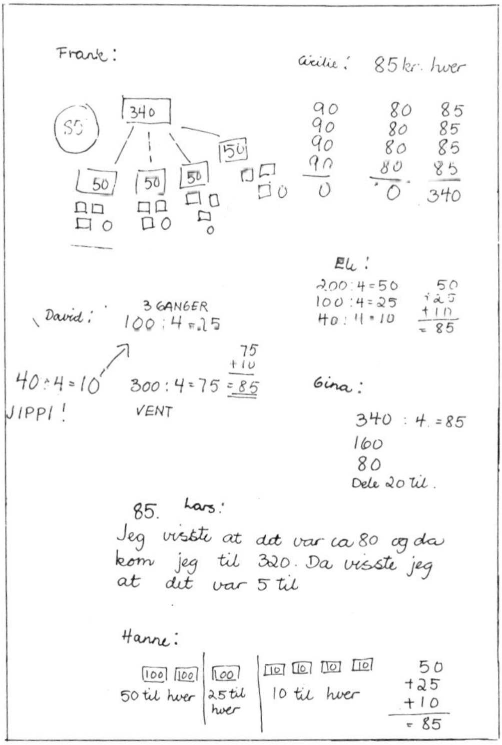

Fig. 2.3.2.

The problem facing the pupils resulting in the above algorithms is 340 divided by 4. They have not yet learned the algorithm for division, so the situation is a problem for most of them. The solutions show much similarity to those ter Hege (op. cit.) reports: variation, imagination, pictorial representation. An obvious task for teacher training here is to examine the pupils' algorithms: what are the actions, and what is their language? How can these eventually be translated into standard forms?

2.3.4. *Progressive Schematising*

On the basis of the recognition of the pupil's functional algorithmic language, and his choice of algorithmic actions, the teacher can eventually help him to transform his algorithm into the wanted standard form, standard with respect to both actions and language. Frequently the pupil will face a *network* of algorithms, and he is supposed to use standard algorithms in order to disclose the properties of this network. But does it always have to be so?

Freudenthal (1981) calls such learning "increasing progressive schematising", and ter Hege (1983) sees this as a main goal for arithmetic learning. But we are here left with one problem: does such schematising necessarily have to be built within the frames which are already set up as standard? Freudenthal writes:

> Schematising should be seen as a psychological rather than a historical progression. I think that in the mental arithmetic of whole numbers we can fairly well describe schematising as a psychological progression, or rather as a network of possible progressions, where each learner chooses his own path or all are conducted along the same way.
>
> Quite a few textbook writers witness efforts to teach learning the traditional algorithms of column arithmetic of whole numbers by a progression of schematising steps, though I am not sure whether their ideas are supported by actual learning and teaching.
>
> Freudenthal, op. cit., p. 140

We remain with a major problem: "What happens when each learner chooses his own path or when all are conducted along the same way?" This statement refers principally to two different strategies of instruction. Ter Hege (op. cit.) makes the following choice:

> Our conclusion after much experience is that children have few problems if they are allowed to attempt in their own ways of multiplication problems —

but then proceeds in a way which makes it difficult for his reader:

— which at a later stage they will solve algorithmically.

Ibid., p. 32

Apparently ter Hege means by "algorithmically" the same as "use of standard algorithms". In this case the childrens' own methods are not given status as algorithms. The methodology is to set the pupils free, permitting them to invent their own methods, and then guide them towards the standard methods.

There are, however, several problems of classroom methodology here which still remain unsolved, both in theory and in practice. The examples ter Hege provides demonstrate the same phenomenon as that illustrated above: children's invented methods differ from standardised methods with respect to both algorithmic actions and algorithmic language. Multiplying 12 with 5 by doubling is a different procedure from the standard one. The doubling procedure may, furthermore, be expressed by an inadequate system of notations. So where do we start — with the procedure or with the notation? If a pupil works out 5×12 by doubling $(24 + 24 + 12)$, writing the whole lot down by means of some private signs on the back of his left hand, do we discuss his method (doubling) or do we help him to write it all down in some proper way? Can we perhaps guide him straight to the standard, written in a standard way, by using spoken language as a means for translation? Where is *progressive schematising*, as the Dutch educators define it, located: in the child's arithmetic (which definitely consists of a system of algorithms), in the standard arithmetic, or in the relationship between these two systems: the continuous struggle of the child to move towards the ideal on the basis of his own functional language?

Theoretically this problem contains a double bind, which implies that in practice it cannot be completely solved. I shall analyse the concept of a *double bind* in more detail in §4.3. But the present problem illustrates the principle perfectly. A double bind rests on the presence of a contradiction. In this case we have the existence of the child's private algorithms taken together with the attitude that the child should be encouraged by his teacher to employ them. Simultaneously we have the presence of a standard system into which the child must be socialised. It is this contradiction between "private" and "social" which causes the tension, and which creates the double bind. There are two keys to obtaining as good a solution to this problem as possible:

A. Helping the child to a metaknowledge about the contradiction causing the double bind, such that he “knows” that his functional language exists in relation to some standard language.

B. Finding an appropriate level for control by means of the standard language. If this control is exercised on too low a level, the child will not realise his own potential of developing and inventing a functional language. If it is activated too late, on too high a level, the child may find himself lost along a blind alley, locked in by a language which is functional only for himself.

We can foresee exciting research in the future here. The first major step has already been taken, the one freeing pupils from the burden of the absolute use of standard methods which prevents their having a functional relationship to the languages of mathematics. Several problems of socialisation remain. I have made attempts to locate some of them above.

CHAPTER 3

LEARNING FROM ANTHROPOLOGY

3.1 RELATIONSHIPS BETWEEN VERBAL AND NON-VERBAL MATHEMATICS

3.1.1. *The Importance of Non-verbal Mathematics*

So far I have paid much attention to mathematics as a language. I shall also examine various aspects of language use as means for the learning of mathematics in what follows. This should not, however, be interpreted as an attempt by me to reduce mathematics to language. I support the view of Otte (1984) that algorithmic thinking (Algoritmus) and visualisation are complementary and corresponding elements which are inseparably located at the centre of mathematics. This position, taken together with the paradigm that language is culturally dependent, implies that visualisation must similarly be culturally dependent.

Extended to the framework of Activity theory, which argues that a functional language has to be related to the Activity of the individual, we obtain the following corollary: in order to understand and promote the visualisations of the individual, the educator has to understand her Activities.

The non-verbal aspects of mathematics learning are so important that they have to be considered. Clements (1982) refers to names such as Einstein, Faraday, Maxwell and a few more to underline the argument that visual thinking is important for mathematics. He refers to Hadamard who insisted that words were totally absent from his mind when he really thought, and that they remained absolutely absent until he came to the moment of communicating the results in written or oral form.

Freudenthal (1978) has much to say about the present situation in geometry education. He refers to pupils who say that they "can see it, but not say it", and argues for a geometry which permits the pupil to talk more informally about what they ca see than has usually been the case. He introduces the notion of *condensation kernels* which can make the internal vision externally visible. He also mentions a few examples of such kernels:

— Pupils explain to each other what they see
— Pupils are being asked to draw what they see, thus making their visualisations visible.

Freudenthal argues for a geometry for children which helps them to experience and interpret the space in which they live, breath and move. Otte (1983a, 1984) for his part explores the role of visualisation for formal mathematics. He stresses the significance of metaphors for the understanding of theoretical knowledge. He considers visualisations as means for such metaphors; even the mathematician with her interest in the formal aspects of mathematics uses "ideographic visualisations" as metaphoric means to develop the mathematical models in their formal form (Otte 1984).

We use graphs and diagrams of various sorts to understand theory and in order to build theory. From the perspective of Activity theory we now face many problems and exciting and promising fields of research. To mention a few:

A. What is the nature of the dialectics between "what we see" and "what can we say about it"?

What are the conditions for a transition of "what we see" into "what can we say about it"; how, where and when do such transitions take place?

Will language determine what we see, even if we cannot talk about it? I shall discuss this latter problem in §3.2. when analysing the Sapir-Whorf hypothesis.

B. What about a situation where the pupils belong to a culture x different from culture y as regards language and visualisations, and the mathematics to be learned obviously belongs to culture y?

What will the pattern of translation from x to y be (see §2.2.), and can we say anything at all about when such translations are desirable?

C. To what extent will an ideological superstructure of language develop corresponding to ideologically biased mathematisations? I shall discuss this problem in Chapter 5.

3.1.2. *Research about Non-verbal Mathematics*

An increasing amount of research on non-verbal mathematics has been carried out over the last decades. Bishop (1980), Lean and Clements

(1981) and Clements (1982, 1983) all provide excellent reviews of research about spatial ability, visual imagery and mathematical performance. The picture they paint is clear at a distance: the non-verbal aspects of mathematics learning are so important that they have to be taken into account. As we approach the details, we find the state of affairs somewhat more confusing and we are frequently confronted with conclusions which are too diverse to provide much guidance for the educator.

On the basis of his laborious analysis Clements (1983) manages to reach a position from which he can make some helpful and important conclusions. He stresses that many highly original and significant creations of the human mind have largely been the result of nonverbal mental representations (mainly visual imagery). This the educator should obviously note. Furthermore, he stresses that mathematics educators need to develop instruments for assessing the role of visual imagery in mathematics learning. At present the psychological dimension of imagery has not been established, and this prevents the understanding of imagery and its role in education.[1]

One problem which Clements faces is the lack of a unifying theory in the field. He reports several competing theories, each of which builds on a specific conception of the role of imagery for cognition and perception. The same seems to be the case for the relationships between spatial skills and the learning of mathematics. Fennema (1979) stresses that even less is known about the effect that differential spatial visualisation skills have on the mathematical learning of females as compared with males.

Clements (op. cit.) seems to stick to the visualisation/verbalisation hypothesis, which claims that there may exist individuals who are basically visualisers, verbalisers, or correspondingly, geometric types, analytic types and harmonic types (Krutetskii 1976).

But here I become somewhat worried. In the consideration of the relationships between mathematical performance and spatial abilities I find no information about the kind of mathematics curriculum which is being tested. If this curriculum is not interpreted as a variable, I am not at all surprised that the empirical research points in different directions. In this context it is also surprising that even Fennema (op. cit.), who is one of the front-runners concerning girls in and into mathematics, apparently does not show any concern for the content of the curriculum content. Although she writes

> One indication that spatial visualization is an important consideration in the concurrent development of sex related differences in favor of males in mathematics achievement and spatial skills

she does not manoeuvre herself into a position where she can focus on reasons for such sex-related achievements:

> However the Fenemma-Sherman studies (16) specifically investigated the relationships between mathematics achievement and spatial visualization skills and these data do not support the idea that spatial visualization is helpful in explaining sex-related differences in mathematics achievement.
>
> Ibid., p. 393[2]

It seems that much of the research on visualisation and spatial ability has been carried out in the tradition of cognitive psychology rather than the tradition of cognitive and social anthropology. That is, the research was mainly worked out by means of series of test situations (as in the case of "differential visualisation skills") where the individuals were confronted with various pictorial representations and asked what they could see, or to manipulate them in various ways. To the extent that this has been the case it is hard to see how such research can contribute constructively to a theory of mathematics education.

First of all, we know very little about the "distance" between the configurations to be studied on the various tests employed and the most used geometrical models in a curriculum. This is also a question of context: in an elementary geometric curriculum the models will have roots in a certain history and culture, and they will usually be applied as models of the physical environment of the pupils. How this may be done is worth a discussion on its own. For the time being it suffices to remember that the construction of a curriculum will have such a basis. The basis of construction of a cognitive test will usually be different, as it is related to a theory which traditionally has paid little attention to the factors mentioned above.

There are also reasons to ask which kinds of context the individuals in such an experimental sample will relate to the tasks with which they are confronted, and how does this interpretation of context influence their performance on the test? In other words: what kind of Activity does the individual participate in when participating in the research?

There exists an increasing amount of anthropological research demonstrating the impact of culture on visualisation. I shall examine a recent strong case of such research (Pinxten 1983) in §3.3. I shall, however, take advantage of Pinxten's work to argue for the necessity of not restricting such an analysis to the frames of cognitive anthropology as he does. Activity theory will demand the methods of social anthropology.

3.1.3. *Language and Culture*

Very few would today argue against the importance of sociolinguistics for education. The findings of researchers such as Bernstein (1975), Halliday (1978) and Labov (1972) have convincingly demonstrated that the language of school is frequently culturally biased in such a way that it causes negative attitudes towards school learning among large groups of pupils. These attitudes occur because of both social oppression by a dominant culture, and the discrepancies between various ways of representing knowledge. The presence of Folk Mathematics (§1.1.2.) provides evidence of such discrepancies.

The conflict manifests itself in its most dramatic form in the case of Third World education (not to talk of Fourth World, which is Pinxten's case), when a curriculum developed in a highly industrialised country is introduced. Austin and Howson (1979) report Strevens, who listed the following key issues for mathematics teaching in Anglophone developing countries:

> 1. Do the teacher and learner share the same (first) language?
> 2. Do the teacher and the learner share the same culture?
> 3. Do the teacher and learner share the same logic and reasoning system? (And is this the logic and reasoning we find reflected in mathematics?)
> 4. Is there a "match" between the language culture and logic/reasoning system of pupil and teacher?
>
> Ibid., p. 162

Austin and Howson are furthermore led by their investigations to consider the following issues, which they claim have relevance far outside Anglophone Africa:

> (i) The language of the learner: the way in which his developing mastery of it is influenced by and influences his learning of mathematics, his thought patterns and his formation of concepts; its "distance" from the language in which he is asked to work mathematically, and from that language which has gradually evolved and now possesses international currency as the language of mathematics;

(ii) the language of the teacher (and author): its use and selection, both in oral and written form;

(iii) the language(s) of mathematics: its similarities with and differences from vernaculars; its development, richness and potential.

Ibid., p. 163

We thus face a complex field of research, where little has been done so far. On the language side we have the relationships between the language of the learner and that represented by education, in terms of status, function, vocabulary and grammar. On the other side we have the emerging experience of the significance of the cultural influence of non-verbal mathematics. Finally, we have the paradigm about the complementarity of the non-verbal and the verbal aspects of mathematics, as inseparable elements of the same field of knowledge. A series of problems about difficult relationships is gradually unveiled.

Things were simpler before. Not so many people had to learn mathematics.

3.2. THE SAPIR-WHORF HYPOTHESIS

3.2.1. *A Language Dominance over Thinking?*

One of the most appealing and widely discussed sentences we meet when we investigate linguistics for the purpose of education appears in socio-linguistics. It is the Sapir-Whorf hypothesis (S-W) which claims that the way the individuals of a society interpret reality is determined by its language:[1]

> We are thus confronted to a new principle of reality, which holds that all observers are not led by the same physical evidence to the same picture of the universe, unless their linguistic backgrounds are similar, or can in some way be calibrated.
>
> Whorf 1967, p. v

One has to be intrigued by this thesis, as it obviously contains some seeds of truth; at the same time it just has to be wrong if it is interpreted by the letter. Its validity is recognised when it comes to simple classifications: English has one word for snow, Eskimos have a variety of words, covering snow falling, snow on the ground, drifting snow and so forth (Sampson 1980). And as I have indicated, it has been a difficult problem while writing this book, that whereas German and Scandinavian can offer "virksomhet" besides "aktivitet", I have to write

Activity to specify the use of the term in English. Of the two researchers (S and W) Sapir was the more modest about the range of validity of the thesis. He was also open to a dialectic between language and cognition, and tended to restrict the impact of language to its vocabulary. Whorf for his part was much more drastic. He included the significance of grammar as well, and even hinted that language would determine logic (Schaff 1973, Sampson op. cit.). Whorf's position was thus that all aspects which could be connected with the term "language" coexisted with the environment. This view clearly comes through in Whorf's works on Hopi, a tribe in Arizona. This language is "timeless". Whorf, originally an engineer, speculates on what a "timeless" physics would be like. He introduces *I* for intensity, suggesting that every thing and event would have an *I*, or be a function of *I*:

> A scientist from another culture that used time and velocity would have great difficulty in getting us to understand these concepts. We should talk about the intensity of a chemical reaction; he would speak of its velocity or its rate, which words we should at first think were simply words for intensity in his language. Likewise he at first would think that intensity was simply our own word for velocity. At first we should agree, later we should begin to disagree, and it might dawn upon both sides that different systems of rationalization were being used.
>
> Whorf, op. cit., p. 218

So, according to S-W there exists a language tyranny over cognition and perception.

It is at this point that the choir of protests joins in, because such a tyranny cannot be total. Sampson (op. cit.) uses the example of Einstein, whose new account of the "grand generalisations" of physics seems as fully alien from the standpoint of received views as the Hopi approach, and Einstein spoke only what Whorf would call a Standard European Language. Sampson concludes:

> Rather than saying that if the Hopi had developed physics then physics would have looked very different, it might be more appropriate to say that if the Hopi had developed physics then the Hopi world-view would have changed.
>
> Ibid., p. 88

I would put this another way: The Hopis have survived through generations without a concept of time. They have coped with nature, production and each other. Thus their Activities have not needed time as a concept. This does not of course imply that they want to cope with reality without a time concept forever. Material conditions can change.

The influence of the "time" cultures may be so strong that the need for the concept enforces itself. There may grow political desires within the tribe itself, influencing culture and production in such a way that the concept is required. Language is no autonomous system which develops independently of other human systems. It cannot be the only vehicle for its own development.

There are apparently two major restrictions related to the S-W hypothesis which demonstrate its limitations. Both are of interest to us as mathematics educators.

In the first place, as Bruner (1974) stresses, the S-W hypothesis is restricted to the lexical level of a language, as a coding system of events: some languages code certain domains of experience in more detail than others. The research which has been carried out to investigate the hypothesis has mostly dealt with the vocabulary at a single level of generality, that is, the words of the language rather than its structural relationships.

The most popular field for such research has been the connection between colour discrimination and the classification of colours in a language. As the human being can discriminate far more colours than a language can have a vocabulary for (Sampson, op. cit., mentions 7,500,000 as a figure), it is a problem for any language to group colours by means of labels. According to the S-W hypothesis individuals should classify colours according to their language, and languages with different classifications should thereby impose different discriminations on their users.

So researchers have set up various ingenious experimental designs, where bilingual and monolingual individuals have been confronted with colours and asked what they could see. The hypothesis is usually confirmed by such research. But as Bruner (op. cit.) stresses, as soon as relational uses of such first order concepts as *colour* are included in the experiments the hypothesis falls to pieces.

(I still cannot but wonder how "bilingualism" is usually generally defined without any concern about how the second language functions in comparision with the first. See §2.1.8.)

Bruner concludes:

> In summary, it appears not as a criterion from our own and other work that linguistic encoding of the stimuli relevant to a given problem can affect the ordering of stimuli by providing a formula for relating them across time or space.
>
> Bruner (1974), p. 383

The other limitation of the S-W hypothesis is that it is based on a rather descriptive view of linguistics. Although Sapir, in some of his writings, stresses that in order to understand how man uses his language, one has to ask him about it, both he and Whorf describe language use at the surface level. We do not hear about the individual's rationale for his use of it. Concepts such as *motivation, intention, need, goal,* do not exist, and nor do contextual elements of the situations in which language is observed.

Sampson (op. cit.) makes a point of this when discussing the relationships between logic and language. Logic is formally connected with language: logical predicates, logical calculus and so forth have emerged as superstructures for thinking through language. But man's thinking is not enslaved by formal logic.[2] Sampson's case is the proposition "P and not P". Is it false? If P stands for "I want a dinner", my weight is 120 kg and it is 3 hours since I had my last meal, there can be some possibility that the above proposition contains some truth for me.

The lack of consideration of contextual factors also seems to be present in research about colour discrimination. If an individual can discriminate between colour A and colour B through her language, there are certainly reasons for this, reasons which are connected with history, tradition, production – Activity. When the researcher enters the life of this individual and does this in the positivistic tradition of research, a possible rationale for discrimination will be that the researcher or the assistant of the researcher seems to be a nice person ("It cannot do any harm to do as she says — she also says I will get 10 kroner for this and that is the best money I ever earned").[3]

In these situations there is a possibility that the individual will discriminate, by applying her language in a traditional and static way, that is, as she has learned to use her language, without looking for the possibilities to develop, transform and extend its range. On the other hand, if some queer ants intrude the village and attack houses of colour $(a + b)/2$ and not houses of colours a or b, I would invest quite a lot in the prophecy that the language of the village would soon specify this "new" colour in its language.

We all have our needs, goals and outlook on our world which provide us with rationales for our behaviour. In this way new categorisations, new relationships and new world outlooks develop: Activity leads to new meaning and new Activity. At the same time new words and expressions may be constructed.

Sampson concludes about the S-W hypothesis:

> When they suggest that we are the helpless prisoners of the categorization scheme implied by our language, Sapir and Whorf underestimate the ability that individual men possess to break the conceptual fetters which other men have forged.
>
> Ibid., p. 102

I close my eyes to his use of "individual".

3.2.2. *Language Development as Part of Activity*

Bruner and Sampson conclude that the S-W hypothesis is probably most valid when it is trivial: language organises simple classifications, not relational thinking. We thus face many problems as mathematics educators. What if we want to communicate some mathematics based on certain relationships, and the language of the pupils does not contain the linguistic expressions for these?

Evidence about such situations is provided from Papua New Guinea (PNG). Bishop (1979) reports how J. Jones (1974) asked local interpreters of PNG languages to translate some mathematics tests into the local language. Many questions were impossible or very difficult to translate, As examples, Bishop mentions the lack of comparative constructions: "*A* runs faster than *B*" has to be translated into "*A* runs fast; *B* runs slow".

P. L. Jones (1982) gets quite close to the problem by his research on their use of relational terms "more" and "less". He stresses that the use often depends critically on the context in which they are used. As an example, he mentions "five is more than three" and "five more than three", the former to be translated into $5 > 3$ and the latter into $5 + 3$ or $3 + 5$. Subtle changes in context can thus dramatically change the mathematical meaning and lead to errors of understanding if not recognised by the learner.

PNG children will particularly suffer from this as English is the language of instruction. Jones (1974) finds that the PNG children make qualitatively the same errors as English children, but quantitatively lag some four years behind. The deficiency of an appropriate relational language is, however, not a complete barrier for the learner. The educator must obviously help the learner to develop the appropriate relational language before the mathematisations begin. According to our Activity theory, the educator has to consider the *functionality* of this new relational language as experienced by the learner.

This problem is deep and difficult for the educator and not yet solved, as far as I know. One promising approach has been made by Pinxten (op. cit.). It is not directly in the field of relational language, but rather with respect to the conceptions of space. In our discussion of Pinxten's work the S-W hypothesis will work in the wings. It will be there as a thinking-tool for handling the status quo of the everyday life of a culture. It will be useless for interpreting the dynamic development of this culture as it grows and produces new ways for its members to coexist and survive. That is, it is useless for interpreting the dynamics of Activity.

3.3. FROM ONE CULTURE TO ANOTHER

3.3.1. *Not Only Euclid*

First of all, a word of warning. My reader should not be deceived by the dominating presence of Fourth World cultures in the analysis below. It is my conviction that many of the problems which will be discussed are more relevant for Western classrooms than we tend to like. I shall be specific about this in §3.3.5.

Bishop (1979) reports about Paiela space (the Paiela are a PNG people):

> (i) it is not a container whose content are objects. It is a dimension of quality of the objects themselves, as their locus.
>
> (ii) space is a system of points or coordinates as the loci of objects. Objects are defined through binary opposition, as large or small, long or short, light-coloured or dark-coloured; and space as the coordinates of objects so defined, becomes axial rather than three dimensional, as up or down, over there or here, far or near, and so on.
>
> (iii) space is not objective but the product of the observer's perception of opposition in sensory data. Among other things it means that size (for them) would be like value (for us), not absolute or gauged by objective measures, but relatively dependent upon the subjective factors of evaluation and scale of comparison.
>
> Ibid., p. 143

We foresee some problems if we confront unprepared pupils of the Paiela population with Euclid in three dimensions. The same would be the case when facing the Navajos, an Indian people of North America. Their spatial conceptions have been studied by Pinxten and his collaborators (Pinxten et al. 1983). Their analysis is probably the most profound and far-reaching in this field, as they not only set out to analyse the conception of the space of a culture foreign to themselves, but also provide a theoretical framework for similar research in other cultures.

Very briefly, the Navajo world is seen and conceived as a combination of two dishlike phenomena. Earth is seen as a dishlike form, hanging upside down. Heaven tops the earth, without touching it. But above all, Navajo concepts are dynamic. The earth is expanding in a gradual movement — changing. In the Navajo world nothing is ever the same. Any phenomenon of the Navajo world will at some point or other be subject to movement. Thus lines are recognised as forms of paths or directions for movement.

Let these only be read as outlines of the Navajo world. The reader is recommended to go to the original source. As an example of cultural relativism, the study is an ideal example for all involved in curriculum construction. The ingenuity, consistency and depth of the Navajo geometry are also important to any reader of the First and Second World who may find it painful to discover that her own geometry is only one among several.

This point also becomes clear from the works of Harris (1980), who discusses the knowledge culture in tribal Aboriginal communities. These have not usually been recognised by Western societies as the most advanced in the world. Harris points out that the Aboriginals knew North, South, East and West before the white man had thought of the compass. Furthermore, with some justification, Harris wonders about what kind of geometry led to the construction of the boomerang. I shall now consider some implications for education based on Pinxten's discussions. Although his theme is Navajo geometry, his discussion has a generality far beyond this. We have to look for its possible implications within what is commonly regarded as comprehensive cultures as well, such as the British and Scandinavian cultures as well.

3.3.2. *Pinxten's Solutions*

The situation is somewhat like this: On the one hand we have what Austin and Howson (op, cit.) say has international currency as the language of mathematics, that is, the mathematics usually presented through Western textbooks. This is based on certain conceptions of space, time, economy production, transport, use of leisure time and so forth. In short: Activities typical of people who have learned to cope with reality in modern, highly industrialised countries.

On the other hand, there are other cultures where Activities are quite different. They will often be Activities for life in societies which have not reached high levels of industrialisation. Sometimes, as we have seen, we may find other conceptions of space, time and production, and thereby also other foundations for mathematics learning. Imposing Western curriculum onto a Navajo culture would imply, as Pinxten stresses, a schizoid situation, both for the teacher and her pupils: there would be some knowledge from the Navajo culture, some " woefully incomplete" knowledge from the Western tradition, and problems for both teacher and pupils of comparing, transforming, shifting and translating knowledge from one culture to another. So how can this problem of cultural alienation be met by education?

Pinxten discusses three patterns of strategy for a Navajo mathematics, or specifically, a Navajo geometry:

A. Teach the Western system.

B. Elaborate the Navajo system for later integration into Western geometry.

C. Integrate the Western outlook within the Navajo world view and in terms of the Navajo spatial model.

Solution C is in many ways the most promising, if something more than frustration, misunderstanding and negative anxiety is to be achieved. But, as I shall soon argue, it is really not our job as scientists from the West to make suggestions about whether it is best to end up with a Western model or not.

Solution A above is the most common solution in Third and Fourth World mathematics education today. Some of the reasons for this are revealed through discussion of solutions B and C. Mathematics education — and mathematics — has few traditions, if any, for such approaches as indicated by B and C.

B is a complicated solution, and Pinxten himself sees clearly the kind of trouble he faces. It implies developing the Navajo geometry for amalgamation with Western geometry. As Navajo spatial distinctions offer genuine qualitative material for geometry teaching, there might possibly be ways of developing Navajo geometry in the direction of Western geometry. In practice such an approach would amount to the translation of the basic Navajo notions into existing formal Western theory.

The suggestion Pinxten makes is to use Thom's dynamic topology or catastrophe theory which can possibly embed both the spatial notions of Western geometry and those of Navajo space. In a second stage, then, the Western and the Navajo system can be compared within the unifying theory, by translation of premises and statements from one geometry to the other.

Finally, Pinxten suggests that exact adaptation procedures must be devised in order to ease the transition from the Navajo statements to Western geometry.

This approach obviously has the advantage that the transition to Western-biased geometrical notions is understood and taken into account on the basis of the pupil's own geometry. In this way Western notions can, after a while , safely be taken as a point of reference, as these have become familiar to the pupil. Theoretically this approach seems fascinating: finding some general theory which can imbed two geometries so that they can be compared with one another within the frameworks of their generalisations.

I see quite a few problems in applying such an approach for the purpose of classroom instruction, at least in the primary and secondary school, and so does Pinxten. He is also aware of the paradox his approach implies: by finding an existing grand Western theory, into which he can embed the "foreign" geometry, he at one and the same time puts a mark of stigmatisation on the "foreign" as something which has to be controlled, and builds another avenue to Western knowledge and Westernisation.

Let us therefore look at his third suggestion. The solution C implies the integration of Navajo and Western systems into a Navajo biased framework. The argument here is that the explicit treatment of the Navajo spatial knowledge in geometry courses would improve the unhappy schizoid situation which would result from solution A. Pinxten demonstrates how such a procedure could take place. It ties up nicely with what would be an application of Case 2 and 3 of the Høines triangle and the recommended Case 3—Case 2 transition. His example is "volume".

The children play with different types of objects which cover instances of volume encountered in the Navajo culture. Throughout this process, and at many other stages of this explicit exploration of the Navajo world of objects, there will grow a more abstract and global notion that all these phenomena are but instances of "volume". Then particular exercises on "volume" can be introduced. On this basis of a fully developed and well-trained knowledge of the Navajo spatial system, the more or less corresponding Western system can be instructed. Pinxten's position here is what I have argued for in §2.2 If a new language is to be learned on the basis of another, the first language has to be functional.

The Zimbabwe Tower
Mathematics textbooks provided by the First and Second World for Third World education, are worth a chapter on its own. On January 31st 1984 there was an exhibition of textbooks for and from the Third World at the University of London. On the 28 stands I found six schemes of mathematics textbooks. Most of them were so "Western" that I could have used them in Norwegian classes without any changes. The few "local" examples which were included would just have served as exotic problems for the Nordic pupil.

The most interesting scheme was the one published by the Longman group: *New General Mathematics*, A Modern Course for Zimbabwe, by J. Channon et al.

There were four authors with British names and an advisory panel consisting of representatives of the Zimbabwe schools (consultants, inspectors etc.)

The opening of the first volume is promising: "Say the numbers from 1 to 30 in your mother tongue." "What is the base of counting in your mother tongue?" "There are many languages in Zimbabwe. Find the basis of counting in as many languages as you can."

This kind of exercise disappears after these first pages. We are introduced to set theory ("Bread does not belong to the set of vehicles), spatial geometry (What is the shape of a tin of Nespray and Full Cream Powdered Milk?)

Finally, on page 50, we find a picture of the conical tower of Great Zimbabwe. Actually, it is not conical at all:

Fig. 3.3.1.

The tower is extended as below in order to obtain a conical shape.

Fig. 3.3.2.

I read the dotted lines as a question:
We have this Zimbabwe tower. It is not conical. It ought to have been. We do not really know what its extension would have been. The advisory panel has probably disappeared. Or it was loyal to conical shapes.

3.3.3. *Who Is to Decide What Is the Best Solution?*

I can now challenge Pinxten from the perspective of Activity theory. Indirectly in Pinxten's discussions there is an assumption that the scientist has something to say about what the "best" solution will be. At least, he does not state emphatically otherwise and his own considerations do not include the Navajos as decision-makers about the solutions of the problem. I may perhaps do Pinxten an unjustice here, but the point is so important that it has to be stressed.

The choice of a curriculum is not the choice of the Western anthropologist or the educationist. It is a *political* choice. Pinxten argues solely for the one thing or the other in terms of differences in spatial conceptions, cognitive systems, and what stress the pupils and their teachers might face when practising one curriculum or the other. Although he is apparently on the Navajos' side, he reasons within a cognitive anthropology clear of conceptions from social anthropology, that is conceptions about the individuals' rationales for learning. The real choice of a curriculum will ultimately be a political choice, to which the scientist will contribute in one way or another.

The Navajo people probably are under US jurisdiction. Perhaps the advice which Pinxten provides will be granted as recommendations by the US authorities and thus have an effect on a future Navajo curriculum. Still, it is a political decision. It is a decision which says what this particular scientist recommends should be the guiding principle for a new curriculum for the Navajo people. So whether Pinxten says *x* or *y*, or is listened to by people with influence, he has a political role.

In Pinxten's discussions we never hear about the Navajo's opinions. More seriously from the viewpoint of social anthropology, we do not hear about what their reactions were towards the research, and how these reactions influenced the research strategies. This critique is clearly not a critique of cognitive anthropology as such, it defines its own frames of research. It is a critique of cognitive anthropology as a tool for education. To pursue the point somewhat further, let us hypothesise an example which has clear parallels in the real world.

Let us imagine some developing country of the Third World. Try to think about it as not being infiltrated by European educationists and publishers. Think about it as a nation which has a policy for its future, and as being in a political position in which this policy has a reasonable chance of being put into practice. We make further idealisations: its government represents its people in such a way that what the people think is right to do, its government will try to administer and organise.

To sort out all the possibilities this nation confronts, let us merely consider two marginal cases:

A. It can stick to its traditions and original culture, and base its production on farming and scattered population.

B. It can aim at industry and technology in order to obtain the material standards set by Western measures.

So what then about the choice of curriculum? Is it in the hands of the academics or not? Is it the psychologist or the anthropologist who really makes the decisions if their advice is followed? Whether yes or no the result is a political result. Education will be part of the policy whether this policy is A and B. The researcher can, as Pinxten does, offer possibilities and give evidence about pros and cons. He is, however, not in a position to decide what the "best" solution will be in curriculum development.

The political role of education can be more or less explicit, more or obvious, more or less visible. In the case of the Navajo it is perhaps not so easy to see. In the case of (a) the cultural revolution in the People's Republic of China followed by (b) the four modernisations, the role of education was obvious. As a result of (a) there was a reduction in the length of schooling and attempts to gain workers' control over the curriculum in order to reduce formalism. Following (b) there was the introduction of what many would call "Western norms" for educational efficiency.[1]

So rather than reflecting on "what is good for the Navajos" in education, we should leave such a question to the Navajo people itself. It is really their decision, and this decision is related to a much wider context than that embedded in the walls of the educational institutions: it is related to the context of society.[2]

3.3.4. *The Kpelle School Child*

In his writings Pinxten stresses that in the Western-into-the-Navajo outlook all uses of the concept of "volume" in the Navajo culture should be included in the curriculum. Pinxten thus does not treat "volume" in the wide sense of this knowledge. In which fields is "volume" important for the Navajos? In which field is this knowledge activated as functional knowledge? In what kind of Activities is this knowledge used, and what does the Navajo think about this knowledge and its use?

Is "volume" most important in the context of wood for the winter, or of the containers for fuel? What are the aspirations of the Navajos for today and tomorrow? What are the needs for knowledge in general and for "volume" in particular according to these aspirations? What is the future perspective of their Activities in terms of their present and past Activities, and what will the implications be for their curriculum?

Pinxten is not alone in neglecting the uses of the concepts he studies. We find something of the same when studying the Indigenous Mathematics Project, as reported by Lancy (1978). This project represents an immense effort in the field of cognitive anthropology to prepare ground for a mathematics education in Papua New Guinea. There are more than 700 distinct languages in the country, and the report (ibid.) tells us that at the time they had complete data of the counting systems from 250 languages. Although it is mentioned in the Introduction that the uses of these counting systems were of interest for the development of the curriculum, the remaining report consists mainly of *descriptions* of these systems. From the report we can neither discover for what purposes the various counting systems were developed, nor in what kind of situations they would be either sufficient or insufficient for the PNG society of today.

However, by reading the report we can imagine the mathematics educators of PNG comparing, comprehending and developing these systems according to patterns like those Pinxten draws, without taking into account the kind of Activities which make the new arithmetic language functional.

Gay and Cole (1967) come much closer to the claims I make here, when they analyse the new mathematics in relation to the old culture in the case of the Kpelle people of Liberia. In their book, which is about mathematics education, we find sections about the African context: life in the forest, the village, agriculture, specialists, politics, the secret societies, marriage and divorce, health and tribal organisation.

And even as step beyond this: considerations about knowledge, authority and the learner. The Kpelle school child

> does not pattern the words he hears, nor does he think of mathematics in terms of laws and regularities. Instead, he accepts each item of knowledge as an isolated gem, connected in some mysterious way to the wisdom of an accepted authority.
>
> We see the lack of analysis, this unquestioning acceptance of authority, as the primary stumbling block to the Kpelle child's progress in school.
>
> For him the world remains a mystery to be accepted on authority, not a complex pattern of comprehensible regularities.
>
> Ibid., pp. 93—94

This is the Kpelle story, and it is perhaps a story about Kpelle politics.

There will be similar stories about the relationships between knowledge, power and the pupils from cultures. It is a question of rooting tomorrow's knowledge in the knowledge of yesterday or future Activities in former, and that includes also the examination of such relationships as those Gay and Cole identify here.

Gay and Cole find their solution to the Kpelle problem in the teacher's helping her pupil to overcome his difficulty, to break through the authority structures, by referring all the time to the experiences of the Kpelle children as part of the knowledge. In order to achieve this, their message is one which Mao never ceased to put forward to his cadres: the teacher should let the children be her teachers, gathering this information as a basis for their own organisation of experience in a framework — in our case — of mathematics.

Gay and Cole were not in a position at that point to show in more detail how to proceed in the classroom. Pinxten has done this. He, for his part neglects the importance of learning from the pupils not only about their conceptions, but also how they use their knowledge. In order to push the argument a little further I shall provide some examples from Norwegian Folk mathematics which we have exploited as material for the learning of mathematics.

3.3.5. *Visualisation and Activity: Some Examples*

Knitting. A situation most Norwegian girls sooner or later will be familiar with is this: She has space time, a long journey or a long chat with friends. She has some knitting wool left over, say blue and red in two balls.

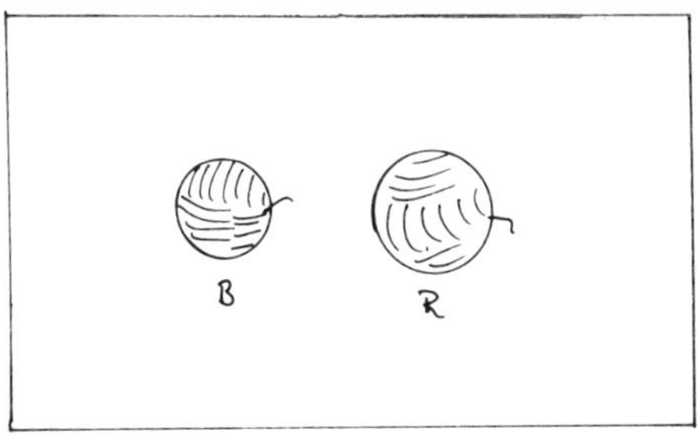

Fig. 3.3.3.

So what should she make? Socks? Mittens? A scarf or a belt? The decision depends on several matters: the quality and quantity of the wool, the time of the year, and what she feels like making. She decides on socks. By experience she knows that she needs 100 gram for a pair of socks, and that this corresponds to a ball of approximately this size:

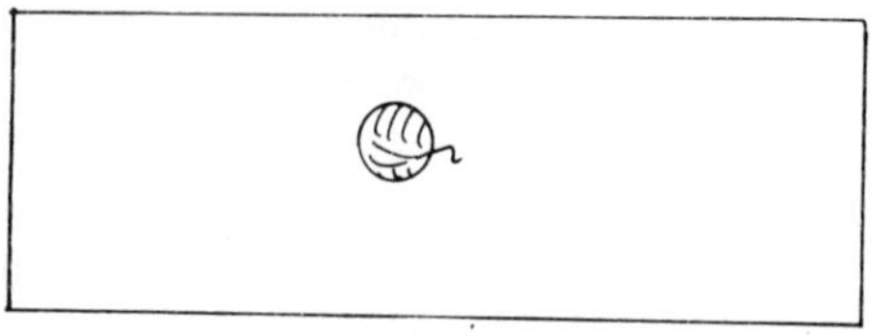

Fig. 3.3.4.

But she has two colours. And people don't usually go around with one red and one blue sock. So she has to design a pattern. She will not draw this pattern. She will visualise it, figuring out something about the ratio of the colours in the various patterns, as compared with the ratio of the quantity of the two sorts of wool she has.

So she decides on a particular pattern. Do not ask me how she does it, she rarely speaks about it. But she has seen quite a few patterns before, and I think that in the end what she fancies depends on her experience. What I know is that she will create new variations of familiar patterns through her work with the wool.

First of all: how much does she have of each sort? That is the first question. She has to divide the wool into halves, as two equal socks are required. So how? The situation might be the one I presented above, with two large balls which are to be transformed into units corresponding to one sock.

Alternatively she has a heap of small balls: which are to add up to her unit balls.

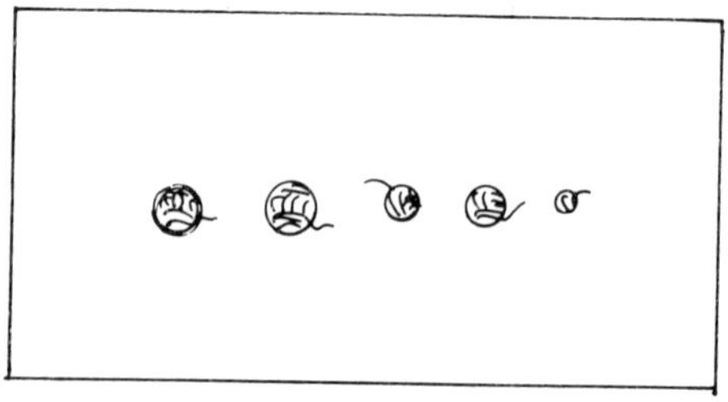

Fig. 3.3.5.

A simple way to solve this problem is to weigh the wool. One sock is 100 gram. Another method which she often follows is to wind the balls into unit balls. Or to make two equal sets of balls. Still, there remains a problem about ratio. If she has 150 gram blue and 250 gram red, she will have wool for two pairs of socks. But what about the pattern? Did we ever see a textbook problem on symmetry which asked about the ratio of the colours involved?

Well, 150 gram is slightly more than half of 250, so for every row of blue she should perhaps have two rows of red. This will be a bit too much red, so sometimes, now and then, only one red before the blue comes up again. The time of the mathematics lesson where the teacher is doing funny things with something called *proportion* is far away.

Rather, she is trying to design a pattern in blue and red where there is almost twice as much red as blue.

Fig. 3.3.6.

The whole lot, including the winding of the wool, takes her less than fifteen minutes. So she starts knitting. While she does so she will discover that her pattern does not work: she has to change it to include some more red (blue) at the expense of the other colour. In the end the completed socks might be one cm shorter or longer than they should have been, but not more. And they are clearly symmetrical as regards patterns and colours.

I have referred to girls: a growing number of Norwegian boys knit; but they still do not knit in public.

Carpentry. More and more Norwegian girls are learning to become carpenters. It is still mostly boys who are observed building in their free time. Boys in Bergen are keen builders. They build for their pigeons, rabbits, and for themselves. It usually starts like this: building materials are discovered, in someone's cellar or garage, on the building site nearby, perhaps somewhere else which I cannot say here. But there are some off-cuts of wood, enough for a hut.

As Bergen is surrounded by hills, and most of the flat land is occupied by the houses of adults, their site will rarely be flat. The situation is somewhat similar to that of the knitter. There exist some materials and some intended end product.

The following procedure is also somewhat similar: there will be some consideration of the design. The planning will refer to the available materials, and in the case of the hut, the site to be used. There will be no drawings, at most the use of an Archimedean rod in the sand. Contrary to the usual situation with the knitting girl there will be several boys present, negotiating and planning in a more or less cooperative atmosphere.

In the case of the boys the visual imagery will be like that of the girl: initially there are no patterns, just the material, the goal, and some experience about how to reach this goal.

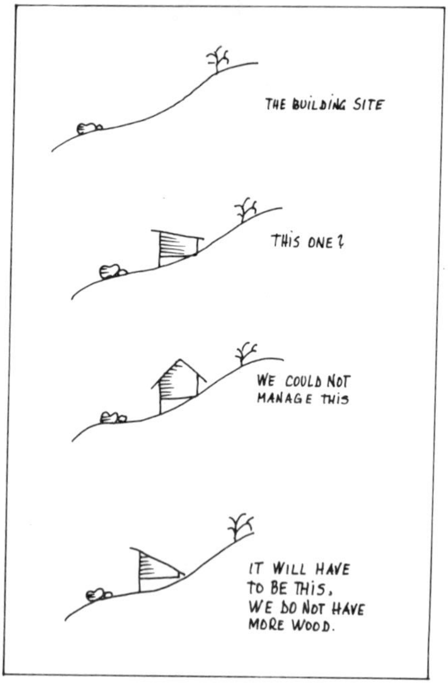

Fig. 3.3.7.

Having discussed and decided on the design of the hut, the builders can start. As in the case of the knitter, the design will change as the boys go on building. It is tempting here to enter into a detailed analysis of how Euclidean geometry is present in a materialised form during the building process, and how this geometry is utilised in other ways than in the traditional curriculum. I shall illustrate the point by two examples:

Traditional Euclid: Two triangles are congruent if their corresponding sides are equal in length.

Carpentry geometry: Whenever you have built a rectangular frame, you need a diagonal bar somewhere to prevent the frame from collapsing. Alternatively: whenever you build a triangle, you ensure that the whole lot does not fall down.

Traditional Euclid: A quadrilateral is a parallelogram if, and only if, both opposite pairs of sides are equal.

Carpentry geometry: You have this frame:

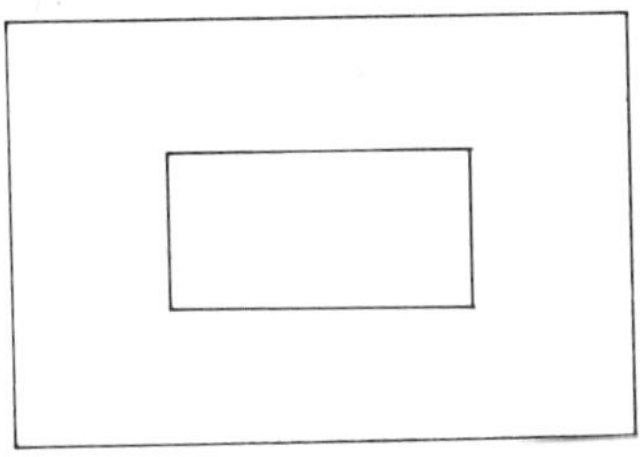

Fig. 3.3.8.

and want to obtain this situation:

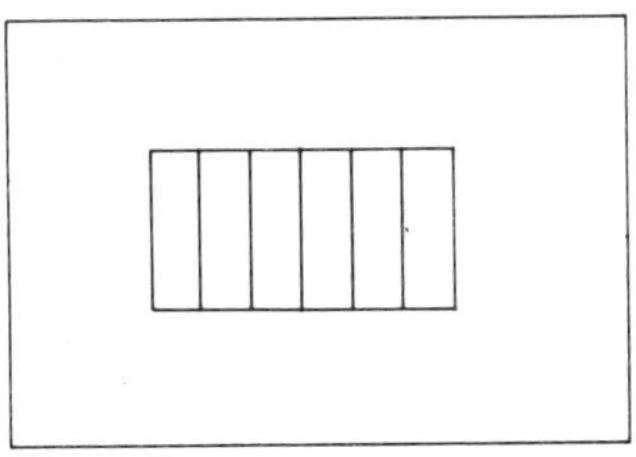

Fig. 3.3.9.

In order to achieve the latter, you measure 60 cm (or less) at a time along both sides:

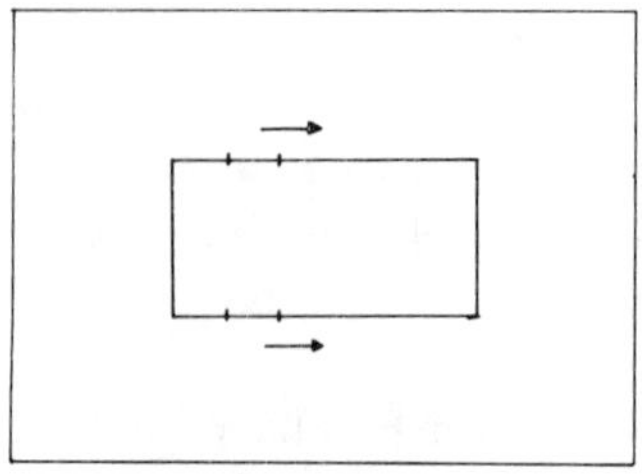

Fig. 3.3.10.

At the end you check by measuring. You then measure the diagonals. You use the theorem that a parallelogram in which the diagonals are equal is a rectangle.

Comments. Many of my readers probably live in places where girls do not usually start knitting when they come across some wool, and boys do not build when they discover some building materials. I am not saying that all Norwegian girls and boys do either. But I have discovered that if I take the time to stay among the young people outside school hours, and do so not only for one evening, but repeatedly, I keep discovering to my mind limitless new qualities of their Activities. It is the method of participating observation I am calling for here. You can observe kids carry out their tasks just by watching them. But when you start listening to their discussions, interjecting a question or two, you realise how well constructed their schemes of knowledge really are. I stress this point because my hypothesis will be that if knitting and building cannot be observed outside school hours, other Activities worthy of exploration within school hours will exist.

Returning to my three dimensions of knowledge as related to Activity (§1.3.), we see that they are easily recognised in the above examples: the historical dimension is obvious. Both knitting and building with timber as craftsmanship have deep roots in Norwegian culture. It is the dream of most Norwegian workers to build their own house, at least a cottage in the countryside from which they originated. This is due to the history of Norway (urbanisation), geography (land available), natural resources (timber available) and politics (a high level of democracy permitting the workers such plans).

The widening-narrowing dimension of knowledge is present. As in the case of the Samba school, there will be a concentration on particular skills, and communication from one generation to the next, from one colleague to the other, of how to perform certain tasks. What school obviously can do in such cases is to focus on knowledge, in its narrow sense, in the context of its use in the children's Activities. I hope the reader detected such possibilities when reading the knitting and building examples. The intrapersonal—interpersonal dimension is most obvious in the case of the boys. They will probably shout at each other, have heated quarrels about how to proceed, and sometimes someone will throw down the hammer and walk away. Still, the end product will be dependent on the interpersonal process.

This process will also be present in the case of the knitting girl. Although she may be the only one knitting in a party or in the train, one is puzzled by the attention she receives from the women around and about the curious questions and remarks made to her. And even if she is sitting alone in her room, reading a book while she finishes her socks, the intrapersonal process still exists in terms of the interpersonal: she has not developed her present skills and experiences in a social vacuum.

The argument so far has been that one has to go further than Pinxten and Lancy did: knowledge cannot, as an object for education, be completely separated from its uses in culture and society. In the case of cultures foreign to the standard school cultures, it is vital to learn not only about their visual imagery and language, but also about the Activities they have developed. In the case of the knowledge culture of Norway I see several important uses of knowledge of which school has usually been ignorant.

In the case of both knitting geometry and carpentry geometry, the utilisation of material is important. It is not a question about "how much do I need?" as we usually face it in the textbook problem. It is rather a question about "how can I use what I have in the best way?"

The relationship between material available and the possible end-result is an important consideration both for the knitter and the builder. And the farmer, the fisherman and the cook. The examination of the end-product by quality control, such as measuring to check, is similarily important. In Euclid we usually reach the result by a minimum of information and operations. In craftmanship one cannot afford the risks this may imply. In the case of the young builders, I might give them a problem like this on their road to Euclid (if this is the goal set for them by the official curriculum):

> You have these materials for the frame of the floor of a hut:
>
> ———————————————— 7.5 cm
> ——————————————————— 9 cm
> ————————————————————— 10 cm
> ———————————— 6 cm
>
> Don't think about what else you will need for the hut.
> How would you make the frame, and how would you check that it is OK?
> Draw what you would like it to look like and write the measures on.

Knowledge is here used in other ways than in standard geometry. One of the few historical cases I know where the principle of bridging people's actual needs and uses of knowledge and the formal systems as represented by school knowledge is the Cultural revolution in China in 1966 and its influence on education. Mao took action to prevent the intellectuals conveying their ideas about knowledge through education. A variety of methods were provided for this purpose: workers would be directors of the schools; worker committees decided on curriculum matters; the teachers had to work in the fields or on the shop-floor for a certain period each year; there was the introduction of barefoot teachers and so forth.

We know that this idyll did not last for long. It is, however, well worth taking a closer look at Mao's philosophy and politics that led to all this. One major message, which I have already stressed, was that the masses in the fields and the factories possessed important knowledge which the intellectuals had to face squarely rather than treat as inferior to their own knowledge.

In his famous speech at the Yenan conference on literature in May 1942 Mao examines the contradiction between popularisation and improving the standard of literature. Mao argues for popularisation as a necessary condition for the improvement of the standard, as the attitude towards literature and art in general among the farmers and soldiers at that time was negative.

For this process to be successful Mao argues furthermore that it is necessary to learn from the masses — "the only source — the largest and richest source, to view experience, study and analyse different people, different classes, different masses, all the vivid patterns in their life and struggle, all the raw material for literature and the arts."

In order to understand the importance of achieving a higher level, the populace has to understand what it is all about — thus popularisation. Or: In order to increase the level, one has to know what basis to build on.

The thrust of Mao's argument here is that they — in this case the artists — have to learn and listen from the activities of the masses. It is in such a context we can regard Hua's presentation at the ICME 4 about popularisation of mathematical methods. Although Hua was severely criticised, since the methods he presented were based on such deep mathematics that one could hardly believe them to be understood by "the people", much of the point is demonstrated by his choice of themes for mathematisation.[3]

Here I shall mention only one: Consider a continuous function with a single maximum over a closed interval. The functional form is unknown and so is the formalisation of the functional relationship. What then is the most efficient way to determine the maximum value with as few experiments as possible? The problem has obvious importance for farmers, the closed interval being the length of the agricultural field. Hua's method, based on repeated use of golden sections, is also easily understood. The mathematics behind it is more difficult, but not more than a pupil in the science form of the grammar school could understand. The point to be made here, however, lies in the formulation of the problem: when did we ever see a grammar-school pupil, or a mathematics student, deal with a function without knowing its formula?

In accordance with such an analysis, I am not convinced that cognitive anthropology, as represented by Pinxten, is functional for educational purposes. I shall pursue this reluctance of mine further in the next section, where I shall take a closer look at Pinxten's theoretical basic, the use of *universals* to determine the cognitive systems of a culture.

3.4. ON UFORS (UNIVERSAL FRAMES OF REFERENCE)

3.4.1. *Old Friends*

Pinxten (op. cit.) introduces the concept of a UFOR as a tool for investigation in the field of the anthropology of space. By doing so he connects it with the philosophy of language and thought. As universalism as a philosophical strand stands in antagonistic opposition to Activity theory I have to take a closer look at it in order to clarify the demarcation line.

My strategy will be as follows: I shall first refer to what Pinxten (ibid.) says about the UFORs. In order to construct my criticism, I shall take a detour to report some of the growing critique of Chomsky, the universalist per se during the last decades. This critique is not directed towards Chomsky's use of universals as a basis for his linguistic theory. It emerges as Chomsky attempts to include another rationale for his theory — the explanation of linguistic behaviour. This critique applies to Pinxten as well: there are few things to hold against him as long as he uses UFORs as tools for determining the anthropology of space in a descriptive way. But as soon as these are applied to the purpose of a dynamic field of practice such as education, it is well worth hesitating in order to take a closer look at the implications.

First of all, some memories which come to mind when reading Chomsky. Has mathematics, as most of us have experienced it, appeared as sets of UFORs? Can many of the arguments we have had over the years about the didactics of mathematics, the very nature of mathematics and its education, be related to/reduced to/examined in the context of the problem of UFORs? Is it here the problems raised in §2.3.4. about progressive schematising are located?

I was always confused when I left his classes. He was a graduate in physics and was tutoring our student teachers in a grammar school. His way of treating mathematical language puzzled me, and his student teachers, who were completely under his authority, even more. Everything was different from what we had discussed at our seminars.

He would draw the graph of some ascending differentiable function and plot in all the familiar notions: $f(x)$, $f(x + \triangle x)$, x, x, use some coloured chalk on the segments $f(x + \triangle x) - f(x)$ and x, by the way, he would call the first difference $\triangle y$ as well. Finally, as you can guess, the fraction and the limit came up, ending with the final statement.

"We call the limit that this fraction has, if this limit exists, the derivative of f in the point x".

Then the circus started. Some pupil was ordered to the blackboard, and had to start from the beginning.

"You call that $\triangle y$. But what was $\triangle y$?"

The pupil points.

"Say it."

"It is that length".

"It is not what I said."

"It's — it's the increase in y — heh?"

"It is still not what I said. Say it."

There was a forest of outstretched hands in the class. They were ignored. It was the teacher, the pupil and the blackboard and the Δy, and the silent moments felt like hours.

"All right. I will help you. Δy is the increase in the value of the function when x increases by Δx."

"Oh yeah".

"Can you now say it."

. . .

"Can someone in class say it now? Ingrid?"

"And Karen?"

"And Otto?"

"Then we can go a step further."

Each step towards the final definition was repeated orally by the teacher and the class three or four times. You can guess how the next lesson would start. And the next and the next.

I never managed to get an attack in. His system was fool-proof. It worked, in the sense that his classes reached the expected standards for the external examinations.

They learned mathematics by saying mathematics. And something must have gone on beneath the surface: they somehow coped with the examination problems. But the student teachers had a hard time. Their tutor was lively and entertaining. He could keep attention easily. Those modest boys and girls from the Norwegian countryside, gifted, but not always expressive characters, they really had a hard time when they had to copy their tutor's demonstration at the blackboard.

3.4.2. *Pinxten's Strategy*

Pinxten's UFOR stands for Universal Frame of Reference. The idea is that the anthropologist should have a research tool which he can use in order to map the various perceptions of space he studies in different cultures. Pinxten stresses that although every culture has its specific way of representing the world, still something is common (universal): it is the very same earth which is referred to, the same sun and the same moon, and all these are experienced by common tools, such as hands, eyes or "by moving around a uniformly structured body in an identical way (e.g., walking forwards and backwards, turning in a horizontal plane), and so on" (ibid., p. 45).

The anthropologist is then confronted with the dilemma that there is, on the one hand, presumably some common basis in the physical and biological conditions for the reflecting man; on the other there exist a variety of world views.

So Pinxten constructs a tool. He makes a list of terms, which can serve as entries into the UFOR. They represent the set of spatially relevant characteristics and distinctions which can represent the world in any culture. He compares it with Roget's Thesaurus: he can match the spatial terms and expressions found in a specific culture with those found in a Western thesaurus. In this way he aims to develop such a tool (which is still not fully developed), whose function is to prepare some common ground between the researcher and the native informant, a neutral and scientific *grid* which represents the problem area under investigation.

I cannot give any detailed description of this UFOR. Just to give a hint what it is about — its first five entries in a provisional list of 345, read:

> 100 Physical or object space
> 101 Spatial aspects
> 102 Near, separate, contiguous
> 103 Part/whole
> 104 Bordering, bounding
>
> Ibid., p. 188

Now to the question of universalism. As hinted above, Pinxten does not assume the position of complete universalism, as Chomsky did on the question of language. How could he, after doing his research on Navajo space? He takes a position which he calls "a posteriori universalism", which implies "first relativism, then universalism".

His point is that there exists an "ontological realism" — that is, there is only one world, one moon and one earth, and man has the same kind of eyes whether he is in Australia or in Finland. On the basis of this assumption he sets out to determine what will be universal, not in order to disturb what he finds in any particular culture, but rather as a large warehouse for his findings. As I have already pointed out: the problem is not the existence of such a warehouse. The problem is the construction of the internal organisational system when educational purposes are implicitly present.

3.4.3. *On Chomsky*

Chomsky is said to have revolutionised linguistics since his first book *Syntactic Structures* appeared in 1957. After the first enthusiasm over his work had calmed down, his method of reasoning has in some way been sorted out and its strengths and weaknesses pointed out. I think that, for the educator, the major problem with Chomsky lies in his conviction of the rationalistic view of language put forward in the seventeenth and eighteenth centuries, which inevitably brings him to the universalistic view of language (Chomsky 1972). A reason for this strand was the dominant position held by Skinner's behaviourism in US psychology at that time (Chomsky 1959).

One point of departure for Chomsky is the observation that the human individual apparently knows more about language than he has learned. Thus it becomes important for Chomsky not only to investigate linguistic structures, but also which predispositions the human individual has for language acquisition. It is thus taken for granted that a person who has acquired knowledge of a language has some internalised system of rules that relate sound and meaning in a particular way. The linguist who constructs a grammar is thus in effect proposing a hypothesis concerning such an internalised system (Chomsky 1972).

The rationalist's view of language assumes there may be some system of propositions in the mind. A received sentence will be realised as a physical signal to this system, and thus given a meaning by this set of propositions. The *grammatical transformations* are the set of formal operations which relate the surface structure of language (the sentence as signals) and the deep structure of propositions.

Chomsky is not primarily interested in such a grammar for a particular language. He seeks the universal grammar, which is characteristic of the human being as a species. He thus tries to formulate the necessary and sufficient conditions that a system must meet to qualify as a potential human language, conditions which are rooted in human language capacity. Thus he can claim to have determined what constitutes the innate organisation which decides what counts as linguistic experience.

It is tempting for a mathematician to describe Chomsky's theory in detail. He is under the strong influence of the stringency of mathematical thinking and borrows his term "generative" for mathematics. His aim was to determine formal means which would generate the grammatical sentences of a language. Chomsky's theory is well described in Chomsky (1972), by Sampson (op. cit.) and in more or less detail in any modern source book of linguistics.

I must, however, stress the implications for *meaning* of the rationalistic view of language. Above I used the expression "generate the grammatical sentences of a language". Implicit in this statement is contained the view that some sentences are grammatical, some are not, and it is the linguistic system as such which tells whether a system is grammatical or not.

For "grammatical" in this context one can read "meaningful", so this view leads to the position that the meaning of a language is not tied to the speaker's knowledge of the world, but is rather determined by the spoken or written sentence itself. The meaning of a sentence will thus be assigned formally on the basis of the syntactic and lexical properties of the sentence per se, and not on the basis of the communicative aspects of the situation in which the sentence is used.

We already see here a complete clash with Activity theory, as communication will be part of an Activity. Before bringing in the relevant critique (still for the educator), we have to bear in mind that Chomsky himself is cautious:

> Evidently, knowledge of language — the internalized system of rules — is only one of the many factors that determine how an utterance will be used or understood in a particular situation. The linguist who is trying to determine what constitutes knowledge of a language — to construct a correct grammar — is studying one fundamental factor involved in performance, but not the only one. This idealization must be kept in mind when one is considering the problem of confirmation of grammars on the basis of empirical evidence.
>
> Chomsky, 1972, p. 27

3.4.4. *Critique of the Paradigm of Universal Grammars*

Olson (1977) makes the point that the controversial issues connected to the debate on Chomsky can be traced back to different assumptions regarding the autonomy of linguistics, according to an assumption whether the meaning of a speech situation is in the sentence per se or not. According to Chomsky each unambiguous or well-formed sentence has one and only one base structure. It is this structure which specifies the meaning of that sentence (its semantic structure). Hence the meaning of a sentence relies on no private referential or contextual interpretations on behalf of the sender or the receiver of the communication. It is from such considerations Olson claims that for Chomsky the meaning is in the sentence per se.[1]

This implies that linguistics as a discipline becomes an autonomous system, and this again causes some problems for the educator who wants to adopt Chomskian ideas as a basis for his curriculum planning.

More recent research on language acquisition has proceeded from an alternative assumption. An utterance is here considered as containing only a fragmentary representation of the intention that lies behind it. Thus the meaning of the utterance is assumed to emerge rather from the shared intentions in the speech situation. The participants' knowledge of the context will here play an important role.

Of this research I would point to Halliday's work (Halliday, 1980). He does not take issue with Chomsky, but the content of his work (as always with Halliday) is in clear opposition to the universal approach. Halliday includes emotions, oppression and class consciousness as context markers for language use:

> The non standard dialects may become languages of opposition and protests, periods of explicit class conflict tend to be characterized by development of such protest languages.
>
> . . .
>
> Here dialect becomes a means of expression of class consciousness and political awareness.
>
> Ibid., p. 80

Halliday is implicitly calling for political considerations when seeking a broad understanding of language use.

It has to do with what one can say and not say, what one can say "in the form of a ghetto language" without outsiders understanding and so forth. Smitherman (1981) discusses Black English, and points out that this language is also

> boastful talk, pungent rhymes, verbal repartee and clever "signifyin" (indirect language used to tease, admonish, or disparage), the rapper establishes himself as a cultural hero solely on the basis of oral performance.
>
> Ibid., p. 45.

The connections of this stress with the political aspects of language use and mathematics education are not as vague as they may appear. If we, as teachers, are to have meaningful encounters with pupils who use language in the ways Halliday and Smitherman here describe (and Labov 1972 as well), we have to obtain as much knowledge about this use as possible. It is this use which can represent the communicative tools of our pupils, and which will mark their demands and knowledge needs thus characterising the educational setting with which we have to cope.

From his conclusion about the universalistic view on the autonomy of linguistics, Olson (op. cit.) ties up his arguments:

> We may conclude then, that the controversy between the syntacticists and the semanticists is reducible to the alternative assumption that language is appropriately represented in terms of sentence meaning or in terms of speaker's meanings. The latter assumption is entirely appropriate, I suggest, for the description of the ordinary oral conversational language, for what I have called utterances.
>
> On the other hand, I propose that Chomsky's theory is not a theory for language generally, but a theory of a particular specialized form of language assured by Luther, and described by the British essayists, and formalized by the logical positivists. It is a model for the structure of autonomous written prose, for what I have called a text.
>
> Ibid., pp. 271—272

As was stressed before, the point is thus not to demolish Chomsky's theoretical framework. The point is rather to make clear its damaging effect as a tool for the educationist. In Sampson's words:

> And certainly nothing in Chomsky's argument for rationalist theory justifies the way in which, for a decade or more, the energies not just of a few enthusiasts but of almost an entire discipline have been diverted away from the task of recording and describing the various facets of the diverse languages of the world, each in its own terms, toward that of fitting every language into a single, sterile formal framework, which often distorts those aspects of a language to which it is at all relevant, and which encourages the practitioners to overlook completely the many aspects of language with which it is not concerned.
>
> Sampson, op. cit., p. 168

3.4.5. *The Berlin-Kay Research*

The motivation for making a case of Chomsky's theory was to learn from linguistic debate the implications for mathematics education in general, and for the UFOR in particular. I can now move closer to the UFOR, by referring to Sampson's (ibid.) critique of Berlin and Kay (1969). They (B-K) set out to verify once and for all that part of the S-W hypothesis which said that the *diversity* of languages basically could threaten the universalist position. Their project was connected with colours, and their aim was to prove the existence of universals of colour discrimination.

For this purpose they used cards of colour chips, the Munsell cards. It was these cards, as used by B-K, which were the inspiration for Pinxten when he started to develop the UFOR (Pinxten et al., op. cit., p. 183).

Now Pinxten is perfectly aware of the strictly linguistic use B-K made of the cards (which colours, how many do you see?). He goes much further in his research approach as he discusses language use with native consultants:

> The complete list of terms and expressions that is presented was worked out, in its entirety, with Frank Harvey and with TB (monolingual consultant). Apart from these older and more eloquent speakers of the language, detailed sets of expressions and terms (ordered sets or just occasional groups of words) were worked upon, discussed, analyzed, compared, illustrated with drawings, and the like by several other consultants in occasional sessions . . .
>
> Ibid., p. 57

The question is: is this enough to overcome the problems caused by the principle of using the UFORs?

The approach of B-K was much more simple, but it illustrates the UFOR principle in an illuminating way. The individuals were confronted with a chart of 127 colours. They were asked to identify the focal colours from this chart. What was the best red, best blue etc. B-K could thus trace how individuals of different languages plotted their focal colours. On the basis of eventual cluster points, they could determine universal notions of colour. One particular sample in the yellow area was chosen as the focal point of a colour term by informants representing eight languages, and its neighbouring samples also scored well. On such a basis B-K identify eleven "smallish" areas on the chart which thus qualify as "universal colours".

I shall refer to another critique which Sampson presents and I shall ask the reader to bear in mind similar research in the field of visualization. One point Sampson makes is that B-K, unlike other researchers in the field, concentrate on the focal points rather than on the boundaries between colours for the purpose of discrimination. Are we interested in the clustering towards some focal example of a notation, or its boundary? When a child calls a hill steep, or a driver calls a bend narrow, are we then interested in what they would call a steep hill or a typical narrow bend?

Or are we equally interested in the domain of hills and bends which qualify for the labels "steep" and "narrow"? The point is made clearer if I add that the child usually refuses to climb steep hills, and the driver is usually prepared to brake when facing a narrow bend. Or even better: are we interested in the focal point of the road engineer's field of narrow bends or the boundary of this field when we see his warning sign on the roadside?

As an educator I am interested in both aspects. Thinking in terms of culture and the importance of learning from its knowledge in order to communicate mathematics, I must know something about the field aspect as well as the focal aspect. The field aspect is necessary in order to get as complete a picture as possible about the use of language. But when teaching some concept we like to use typical examples of it to get its ideas over as efficiently as possible. It is with such examples that we concentrate on knowledge in the narrow sense, in order to generalise and formalise.

The field aspect belongs to knowledge in the wide sense, to Activities, the focal aspect to the focus on knowledge in the narrow sense, as the study of a thinking tool for Activities. Sampson makes another point related to this. He looks at the tradition of the Chomskians of bringing in notations for their universals represented by the focal examples. Sampson's metaphor is the geologist's U- and V-valley. If the geologist happened to bring the U and V as symbols for valleys on their maps, rather than the field oriented use of altitude lines, quite a lot of information would be lost. Sampson mentions the possible W valley which would not be spotted on such a map. The W valleys may be hypothetical, but the point has been made.

So what then is the UFOR? Is it a sample of notations, which taken together, comprise some stance for use in any culture? Is it the notation of any culture to be collected in cooperation with native consultants, poured into that stance, solidified and then used for the purpose of curriculum development?

In this case we foresee a static, sterile use of knowledge, which might be far from the needs of knowledge in the particular culture. Probably Pinxten does not want to be identified with such education. But it is hard to see so far how his conceptual framework and his theoretical basis can prevent him from adopting such a position.

From the position of Activity theory we shall be alert to any system and method of observation which prevents us from feeling the pulse of culture. It is thus that we can learn its basic Activities, and support it — also by mathematics.

3.4.6. *The Claim for Social Anthropology*

Pinxten and his collaborators have, as far as I know, achieved the deepest insight obtained by any European into the knowledge culture of the Third or Fourth World as related to geometry. I myself am convinced that their thinking is also valid for the cultural differences which may be observed within Western societies. I am thereby taking the position that the mathematical curriculum as it is usually practised in Western societies, historically supports mainly the Activities of the privileged classes.

I have argued for the need to go somewhat further than Pinxten has done. I have argued that it is not sufficient to enter a culture without (i) participating in its Activities and (ii) studying how these Activities affect the use of the thinking-tools in general. It is difficult to see how we can learn from a culture basically using native consultants as informers. If some Navajo expert came to Europe in order to learn about European working-class culture and the working-class conception of space, should she then contact the unions and discuss with their representatives? Or should she learn by using informers representing the employers or the educational system? What is the informer representing?

Education — interpersonal processes between teachers and pupils — is dynamic. There is reflection on both sides, and in order to design educational situations in which some significant learning can take place, the educator must know something about this reflection. It provides the basis for the Activities of both teachers and pupils. Social anthropology studies such reflections, or rationales for behaviour. Cognitive anthropology does not.

3.4.7. *Towards a Social Psychology*

A relationship exists between Activities and perception, even the perception of space. And this relationship has to be reflected in the curriculum. Perhaps there exists visual imagery which is not part of Activity. I know quite a few people who can contemplate an Escher piece of art. Perhaps there should be room for the aesthetic experience for purists. I will not take issue about the possible existence of such purism. I do not believe it exists. Activity theory certainly does not include it. Rather, Activity theory faces a much more difficult problem.

Say we reject any kind of frame of reference, not necessarily Universal in the wide sense as Pinxten defines it. Such frames already exist in some form or another. I have already pointed to ter Hege's conception of an algorithm, which apparently says that an algorithm is always of a standard form, as the final result of progressive schematising. Such rigid systems of algorithms, certain ways of solving problems by certain notation systems have some common kernel with the

UFORs. They are universal modes of knowledge which, by means of the curriculum, are supposed to direct the learning of the pupil.

Clements (1980) reports on Newman's and Casey's hierarchies for problem solving.[2] They associate with any given word problem a number of *hurdles* which have to be overcome if a correct solution is to be obtained. The failure of one hurdle prevents a person from progressing to the next. So originally Newman, and later Casey defined a hierarchy of error causes which applies to one step written problems. Again we face an attempt to construct some universal scheme, in this case to trap significant moments in the pupils' problem-solving processes.

A parallel to this is the situation where a child has to ski down a steep slope. I might tell you that there are many different ways of doing this. It is perhaps not irrelevant to say that while some boys are committed to going straight down, some of the girls can relax and climb down the steepest parts. Some of the boys can do neither. The point is, however, that some bright fellow sets up slalom gates which should help all the kids down. These gates are, of course, placed on the most significant parts of the hillside, so the kids can avoid what are expected to be the trouble spots. It is the identification of such spots which everyone must get past which points in the direction of universalism.

The Swede Kilburn does exactly the same thing for the purpose of facilitating the learning of standard arithmetic algorithms. He constructs matrices: in order to perform $A(n)$, you have to master $A(1), \ldots, A(n-1)$. Each $A(n)$ consists of a string of parallel notions for this algorithm. The matrix thus functions as a guide for the teacher to interpret feedback from her pupils.

Such taxonomies as Casey, Kilburn and Newman construct cause the same sort of double bind as the one I described at the end of §2.3. It can hardly be avoided. The taxonomies are helpful for the teacher. At the same time their use is in conflict with the educator's aspiration that the learner ought to experience her own capability by developing her own methods and ways. There will be many ways to get down a steep slope. The one through fixed gates is only one.

It is the *control* of learning which causes double binds. The control of the double bind, by the teacher and ther learner is due to the handling of *metaknowledge* about the control caused by the taxonomies. The structure of a *double bind* is thus essential for a theory of socialisation.

CHAPTER 4

LEARNING FROM PSYCHOLOGY

> *Laws of the class-room:*
> *First law*: Find out what pleases and displeases the teacher.
> *Second law*: Bring to the teacher's attention those things which please the teacher and conceal from him those behaviours which will displease him.
> *Third law*: Remember that it is a *competitive* situation. The pupil must try to please the teacher and avoid displeasing him more than other pupils.
>
> D. H. Hargreaves

4.1. SYMBOLIC INTERACTIONISM

4.1.1. *Understanding Activity*

Previously I have laid the emphasis on the pupils as decision-makers: they decide to participate in teaching situations, and they decide not to. I have referred to pupils who judged their curriculum as sheer nonsense and rejected it. In addition I have described a few projects which have engendered eager participation and interest. I have been looking for means to understand the foundations of the relationships between the individual and the knowledge he is confronted with. In particular, I have been interested in now this relationship is influenced by the social situation of the individual.

On one side we have sociology, which can teach us something about the distributions of economy, ideology, power, etc. On the other side we have psychology, which is the science of human behaviour. Very few would deny that human behaviour is partly determined by the society of which the individual is a member. Still there do not exist many psychological theories which make attempts to explain behaviour in terms of the social environment of the individual.

This was exactly the motivation for giving Activity theory a dominant place in this book. Activity theory is, as I stressed in §1.2.2., about the formation of the social individual as a part of his society.

Activity theory embraces both fields coherently: society and the individual are studied dialectically. Activities are the processes by which the individual exists, copes and survives in his society. The individual is thus a social individual in Activity theory. This issue is not questioned, the label "social" is not used about individuals in Activity theory.

Still, the theory does not say much about *how* behaviour can be interpreted in terms of the environment. Activity belongs to the individual, not the observer. It is exactly this that may be the problem in education: how to interpret the behaviour of others (the pupils) in order to understand their Activities. It is this problem which caused the relationship between the Activity and the educational task analysed in §1.3.: the educationist had to grasp the Activity of the pupil in some way or other and design educational tasks accordingly.

Furthermore, most of us face more complex societies than the psychologists of the young Soviet State did. Activity theory is constructive, positive and optimistic: there is always some goal; there is always some interpersonal process. The only problem seems to be to provide the pupils with some fruitful learning environment which can support their Activities, resulting in further learning and so on.

The reality is somewhat more complex than this. We experience pupils rejecting school. We see kids being destructive, violent. As adults we have problems in understanding their behaviour. It seems appropriate to make some attempt to develop the Activity theory I have built so far in order to obtain some thinking-tools which can help to explain behaviour, both constructive and destructive. The ultimate goal is still the same, to build as consistent a theoretical basis as possible which invites as many categories of pupils as possible to mathematical knowledge. I shall start by exploiting the thinking of the American social psychologist George Herbert Mead.

4.1.2. *The Sociologist's Psychology*

Mead (1934, 1965) defines himself as a social behaviourist. His theory can be regarded as a communication theory, where the various gestures of the individual as communicative means play a central role. His concern is not, as with Vygotsky, the development of gesture systems in the history of man, but rather the interaction process itself which determines the individual in terms of his social environment. Mead's theory is thus a theory of interactionism, based on the symbolic behaviour of the individual. It is, however, somewhat strange to see the label *behaviourism* used in the case of Mead, as he persistently goes behind the surface of behaviour, looking for its rationale. The influence of Skinner in interpreting the notion of *behavourism* narrowly has been strong.

Mead builds a complete psychology. At the centre of his theory we find the human being, as a reflecting and communicating individual, who has a certain logic for this actions. The major conceptual constructs of the theory relate the individual to the social groups in his environment. The individual experiences himself by means of the reactions towards him by the members of these groups. He mirrors himself through others as he notices them signal reactions to his behaviour.

Mead's psychology thus becomes the sociologist's psychology:

> Society not only defines, but also creates psychological reality. The individual *realizes* himself in society, that is, he recognizes his identity in socially defined terms and these definitions become reality as he lives in society.
>
> Berger 1971, p. 108

Mead calls the social environment to which the individual reacts the Generalised Other (GO): it is thus the common attitudes, expectations and reactions as experienced by the individual which constitute the GO, and which function as the individual's referent for his behaviour.

Mead furthermore describes the Self as comprising two related parts, the I and the Me. I is the disorganised and spontaneous part. Me is the controlling and directing part, composed by the GOs which are relevant: family, school, peer groups, etc. Me thus covers the social aspects of the person; it is the individual's representation of society through the attitudes, expectations and meanings of the group.

On this basis Mead defines *thinking* as an inner conversation between the I and the Me. The flavour of Vygotsky is obvious, although Vygotsky did not conceptualise any I in contrast to Me. The Vygotskian *Self* is completely social.

4.1.3. *From the Generalised Other to Ideology*

Clearly the individual is exposed to various GOs. The family and the peer-group are obvious examples of this. The church may be a third; school a fourth.

The analysis of the impact of various GOs on pupils' behaviour is vital in social education. The widespread use of Mead's theory in our time is also due to the unstable life conditions of the pupils. For many there is a rapid shift of influential GOs which school has no experience to handle. It is especially the transition from the influence of the family as a GO to the influence of the peer group which causes concern. An increasing number of children have to take care of their own socialisation by means of their more or less stable peer groups, resulting in other types of knowledge among youngsters than could be observed some decades ago.[2]

The increase in immigration into various countries also demonstrates the value of Mead's theory. On a postgraduate course in social pedagogy, a teacher presented the following case: she had been working with an immigrant child (black) who had become a foster child in Norway after immigration. This child had its original culture as a referent for its behaviour, the new culture, the biological family and the foster family, that is, at least 4 strong GOs.

What we do not find in Mead's theory, and what experience tells us is so influential (as indicated by the above examples), is the situation where several GOs work simultaneously and in contradiction.

What happens to a pupil when school, as a GO, says that education is important, while the family says the opposite? What about the situation where the messages from the peer-group contradict those of the family? In which cases do such contradictions lead to confusion and paralysis and in which cases does the individual manage to sort things out, take his decisions?

Another thing which I cannot see Mead considering is the influence of *communication* between representatives of different GOs. Individual x faces representatives y and z of — say school and social authorities — and y and z communicate (by means of eye contact). In such a case x relates to communication between y and z. What can this mean for the relationship between x and GO y and GO z?

The social formation of an individual is obviously not only dependent on the social control of one particular GO. There usually exists a *network* or a *system* of GOs, and the internal structure of this will be just as important to analyse as the influence of one particular GO. Mead does not provide us with any apparatus for such an analysis. The tools for this will be provided in §4.3., where I shall exploit Bateson's communication theory.

I shall now introduce a concept of *ideology*. I can substitute this term for *GO* without doing much damage to Mead's theory. The attitudes of the people around us, the controlling mechanisms of our environment, what is taken as commonsense knowledge, i.e. the basic ideas of a social group, are all part of an ideology.[3] Some are acceptable, some are not. Some are legal, some are not. Some are kept conscious, some are suppressed. Some knowledge acquires the status of being universal, and thus takes the form of an ideology. All this is what the individual has to relate himself to, in his search for his Self, his personality and his construction of reality.

Marx builds his concept of ideology on a critique of the ruling ideologies of the capitalistic society at his time. These implied a false consciousness by the individual about his social reality. In my use of *ideology* I relate this construct to the individual, in particular the pupil, as a carrier of ideas developed by him in his social relationships, i.e. the attitudes he has adopted from his GOs.

Following Giroux (1981) there is a need to stand back from the major strand of ideology theory employed in education, where most of the work has been done in order to determine the ideological content of textbooks, curricula, frames of schools, etc. When research about ideology on the educational scene is restricted solely to such a field, one will end up in a blind alley, as it leaves the pupil with no possibility of challenging the ideologies represented by the school. In this tradition it is only the educational system which produces ideology, not the pupils themselves.

As Giroux (ibid.) points out: ideology is something more than the reification of consciousness and social relations; it is also consciousness struggling to constitute itself against the objectified nature of social life. Ideology thus not only belongs to institutions, it also belongs to individuals who carry it.

Many of the projects reported in this book are examples of an approach which relates ideology to individuals and challenge ideology by the exploitation of the thinking-tools of the subjects, perhaps leading to some action by the individuals.

A common case today is ideologies related to women's lib. Doing projects on various topics related to female oppression, such as boys' social dominance in the classroom, boys' choices of subjects as compared with those of the girls, possibilities of employment for girls and so on by means of the tools of information theory/mathematics/ statistics will usually challenge ideologies with new insights, and provide some new ammunition for those in need of it.

I shall return to the concept of ideology in more detail in §§4.2. and 5.1. The reader will so far have noticed that I introduced a *dialectical* concept, as ideology's manifestation through the individual is a result of his interaction with his social environment, that is, as a result of the social control exercised by his system of GOs and his own Activities.

I hope the relevance for Activity theory of such a conception is clear to the reader. I have stressed that the concept of Activity is a political concept (§1.2.3.). The above concept of ideology points directly to the heart of the concept of being *political*. It helps us to understand why certain groups of individuals choose the goals of their Activities as they do.

A gang of youths can choose the politics of hooliganism as a basis for their Activities. The gang can choose a different strategy to cope with life. Parts of this choice result from the way the individuals of the gang experience the reactions of their social environment, that is the system of GOs.

4.1.4. *Rationality for Learning*

A basic paradigm for social pedagogy is the position that all human behaviour is intelligent. The problem for the educationist is to understand the basis of the observed behaviour, the kind of rationale on which the behaviour builds. As educationists, psychologists, etc., we can examine the intelligence of behaviour, i.e. the rationale of behaviour, by exploring the relationships between the individual and his material and social environment. The GO's and the individual's ideologies will be part of this.

The individual's rationale belongs to the individual. It is the way he "chooses" to act in his world under the material and social conditions under which he lives. The rationale of behaviour is the result of the ideologies which the individual carries. As I have so far analysed ideology in connection with the relationships between the individual and his GOs, I shall, following Mead, analyse rationale in connection with the relationships between the ideologies of the individual and his behaviour.

This somewhat subtle theoretical model is motivated by the ever-present experience of our pupils: as educators we see them decide to learn or not to learn. As educators we are cheating ourselves if we do not make this phenomenon central to our theorisations. A functional didactic theory of a subject must accordingly have the logic of behaviour as a key structure.

Mead describes such a rationality as

> a certain type of conduct, the type of conduct in which the individual puts himself in the attitude of the whole group to which he belongs.
>
> Mead 1934, p. 334

We notice that the rationale is not the conduct itself, it is a type of conduct, a type which comprises the logic of the individual according to the social setting he is related to. Again I find Mead's reasoning too simple for my purposes as an educator. I have pointed to the fact that there usually exist several groups which influence the behaviour of the individual. "The whole group" is thus not an accurate conception for our purposes, nor is it sufficient. There may be several rationales present for behaviour, as there can be several groups which function as GOs. We can thus observe apparently confusing and irrational behaviour, which results from a system of rationales rooted in correspondingly conflicting GOs.

Like ideology, a rationale for behaviour is a dialectic concept. The rationale belongs to the individual, but is a product of the individual's relation to the system of his GOs. To explain a rationale for a particular behaviour is to describe the system of attitudes of the GOs. On the other hand we have to avoid determinism; the individual is not locked into such a set of rationales. Activity theory considers the individual to be reflecting and acting, thus being in a position to evaluate the effects of his behaviour.

Two major rationales can be identified as drives for school learning. It is the rationale which is related to school's influence on the future of the pupil, by the formal qualifications it can contribute. This role as an *instrument* for the pupil will provide the pupil with an *instrumental rationale* (I-rationale). In its purest form the I-rationale will tell the pupil that he has to learn, because it will pay out in terms of marks, exams, certificates and so forth.

On the other hand there is a rationale which relates to knowledge as such, saying that "this knowledge has a value besides its importance for the external examination." It is a rationale which says that knowledge has an importance beyond its status as school knowledge.

I call this rationale the *S-rationale.* The S stands for "social", indicating that such an evaluation of knowledge is made by the individual as a social subject, that is, by the individual through a reference to his GOs which goes beyond the I-rationale. Obviously it is not the case that either the I or the S will work as a drive for learning. Usually they will work together, and the pupil's rationale for learning may be seen as a resultant of the I and the S: knowledge is regarded important by the pupil for external examinations and for itself.

Whatever the combination of the S- and the I-rationale is the pupils are participating in some Activity. The major problem is when the pupil does not engage in the discussion at all. The vital problem for the teacher is to know *which* Activity. In this case, what is the system of rationales which builds the learning Activity of the pupil? It is not a very constructive situation when the teacher believes his pupils are working with the content out of interest for the subject itself, when what they are really doing is working towards examinations as the major goals. The opposite case may also be worth considering. We are back here to Bauersfeld's (1979) stress on examining "the matter meant"/ "taught"/"learned".

The pupil's evaluation of some knowledge as I- or S-knowledge is clearly not static over time. It will depend on his learning history as well as the historical transformation of his GOs. Some learning based on the I-rationale can imply the activation of the S-rationale for the same type of knowledge. The former I-knowledge may then be transformed into S-knowledge.

In commonsense language: the pupil may work with gradients of functions without having any idea what it is all about. From the examination papers he sees, and, from his teacher's warnings, he knows, that this is important stuff. The I-rationale will obviously be effective for this pupil. Then his teacher demonstrates how the idea of growth can be applied to various increases, things which catch the interest of the pupil (pollution, employment, club capacity etc.) The S-rationale will thus be activated together with the I. Of course, the teacher could have done all this the other way round, it doesn't matter — the point here is to show the non-static nature of the I and the S; they mutually influence each other. This influence works both backwards and forwards: previous knowledge may be related to a new system of rationales, and a prevalent system of rationales may provide new sets of attitudes ("I never saw any sense in gradients, but after that project about pollution, I would like to learn some more about them".)

My reader may now ask why I introduce notions such as "I" and "S" and "rationales" about the commonsense experiences of any teacher. I can now justify that. This simple theory implies some simple corollaries which are rarely considered in educational praxis.

COROLLARY 1. If the I-rationale is sufficiently weakened by negative messages from school, the learning will rest on the activation of the S-rationale.

"School" is here thought of as a GO. School communicates *about* the pupil *to* the pupil in a variety of ways, from the marking system to the suggestive glances of the teacher. For some of the pupils the cumulative effect of such negative messages will sooner or later mount up — enough is enough. They realise that school cannot function for them as an instrument for a proper future. I assume here that the pupils do in fact discover this. With support from research I shall argue that several do (see my Introduction). Our experience is that these pupils decide not to learn any more; they just do not see the sense in doing so.

Still, we regularly face pupils who stubbornly tackle their mathematical problems without understanding what it is all about and without any success, receiving more negative feedback from school, leading to new efforts and so on. This is the double-bind situation. These pupils are, however, not included in Corollary 1 as the corollary assumes a weakened I-rationale. If school does not manage to activate these pupils in Activities based on their S-rationales, it is difficult to see how they can proceed to learn their subjects.

COROLLARY 2. If the pupil stops learning because both the I- and the S-rationale have ceased to function, a remedial education has to build on revitalisation of the S-rationale.

This corollary says that in order to get pupils who have stopped learning going again, we have to provide them with some powerful projects which can make contact with their S-rationales. That is, we must concentrate on knowledge in its wide sense (§1.3.6.), so that the pupils can experience the strength of knowledge as consisting of thinking-tools in contexts outside mathematics itself.

It is amazing how current practices of remedial education neglect such a simple principle of social education. In special education the diagnosis is usually connected with the pupil and not his curriculum. Thus the pupil who is selected for remedial education will frequently be confronted with a curriculum of the same *nature* as the one that he did not cope with originally. That is, the curriculum calls for the same system of rationales as the original did. The efforts of special education will in many of these cases direct themselves towards fields which are remote from the pupil's S-rationale: smaller classes, larger printing in textbooks, simpler problems — but, mind you, the same type of problem as before — concern about a friendly atmosphere etc.

If the roots of the failure in learning are buried in a curriculum which is of no interest to the learner, a strategy of instruction like the one indicated above may *aggravate* the pupils learning problems. It may promote the rationale which is already working which tells the pupil it does not pay to learn *this kind* of subject matter.[4]

4.1.5. *Some Limitations of Social Interactionism*

Clearly the S-rationale is a sociological concept. The difficulty this causes for the teacher is that it will vary within the same class: what will pass as I-knowledge for some pupils will pass as S-knowledge for others. I refer to the discussion in §1.3., and my conclusion that these problems often cannot be completely solved: sometimes the teacher has to choose which group of pupils he wants to favour at the cost of others, that is, whose S-rationales he wants to activate.

Another problem that arises for the teacher, as in most psychological theories, is that the focus is on the individual. Usually he faces groups of pupils. Blumer (1971) makes up for much of this with his concept of a *joint* action. Examples of such actions are a debate, a family dinner, a game, a war. In each case there is an identifying form of joint behaviour which comprises the articulation of the acts of the individual participants. In such an action both the common actions of the participants and those of the individual can be identified:

> By identifying the social act or the joint actions of the participant he is able to orient himself; he has a key to interpret the acts of others and a guide for directing his action with regard to them.
>
> Blumer, ibid., p. 20

The individuals have an impact on the joint act: At a dinner party a dinner-behaviour can be identified, just as the behaviour of an individual guest can be. People participate in the dinner as a party, at the same time as they participate on behalf of themselves: "I never liked that lady they put next to me".

Both *ideology* and *rationale* will be present in a joint action. There will be a certain set of common sense rules for a dinner party. There will be some motivation for participating in the dinner (wedding, birthday etc).

Similarly, there will exist an identity of the school class, or of a group within the class, an identity which the teacher daily faces and to which the various pupils contribute and relate. The pupils will be conscious of being *in* class, and behave accordingly. At the same time they will be conscious of themselves and their own contributions to the class.

So, rather than looking for the rationales of the individual pupil, the teacher may profit by looking for the rationales of the various *groups* of the class. One common situation is the presence of one group whose basic rationale for learning is the examinations and another group which is not motivated by this any more. This situation, familiar to most teachers, is extremely difficult, and can only be satisfactorily dealt with in part. Later I shall argue further for the advantages of planning according to the needs of groups rather than individuals.

For interactionism as it stands, the individual (or group) interacts with other and, this again, with others. It is the study of an infinite chain of dominoes, without a beginning or an end, one piece making the next fall and so on. The initial push is ignored, and so is the final result. As Woods (1983) stresses when he describes symbolic interactionism as a means of investigating the social life of the classroom: at the heart of it is the notion of people as constructors of their own actions and meanings. Symbolic interactionism does not treat the *importance* of events for that is defined by the people under study.

Decisions are not, however, taken out of the blue or solely as a result of interaction with other people. A restaurant dinner where the portions are extremely small, the food is cold and the serving slow will cause a different resulting "joint action" from a perfect setting of the meal.

The destructive behaviour observed among large groups of young people can hardly be described solely in terms of group identification. We should also mention something about the effects of economic depression such as lack of employment, lack of stable social milieux etc., in order to obtain some genuine understanding of the rationales of the group.

Here Activity theory offers itself as a possibility, as it weighs man's potential to transform his reality, by transforming not only himself, but also the world he is a part of. The concept of the S- and I-rationale in a sense satisfies some of the claims made here. They say something about the relationships between school knowledge and the individual in terms of what is important knowledge for the individual. At the same time factors *outside* school are considered: the rationale of the learner as a result of what counts as important knowledge in his social environment for the time being.

Behaviour does not only constitute a dialectic between the individual and his system of GOs. It is also the result of a dialectic between individuals and the material conditions they are exposed to, that is, their outer world. Symbolic interactionism can thus be utilised as a tool for Activity theory, or for the educationist in particular: a tool to interpret behaviour, to see how the various GOs influence and direct behaviour. To use this tool without considering factors outside the chain of individuals can lead to educational strategies which build only on the individual in relation to his GOs, and not the material factors influencing the interaction. A difficult double-bind may be the result of this.

We are getting closer to some difficult and deep problems. I have already several times made the point that Activity theory as developed by Soviet psychology did not conceptualise contradictory world outlooks or different ideas about how to organise Activities as a means of coping with and surviving in the world. Some people are prevented from participating in Activities, and this prevention gives rise to concepts such as hegemony, oppression, power, resistance and several others.

The original adult pupils of Freire were all denied access to Activities. In Western classes the problem is even more difficult: we can face both active and passive pupils in one and the same class. Before suggesting some strategies of approaching such situations, I shall analyse some more theory which I have found helpful.

4.2. PSYCHOANALYSIS

> In a healthy community and in a socially healthy personality, behavioral ideology, founded on the socio-economic basis, is strong and sound, here is no discrepancy between the official and the unofficial unconcious.
>
> V. N. Volosinov

4.2.1. *A Linguistic Perspective*

The reader may ask .himself what a section on psychoanalysis is doing in a book on mathematics education, and he is quite justified. I must admit that I have some hesitation about the heading of the section. There is very little of the original Freud left; and quite a lot of themes from socio- and psycholinguistics will be considered. My major goal in this section is to argue that Activity theory generalises psychoanalysis. This implies that a Utopian vitalisation of Activity theory will lead to the elimination of psychoanalysis.[1]

My project is not only an exercise in some theoretical game. The major message of psychoanalysis is that oppression of behaviour leads to distorted behaviour. Another important message is that oppressed experience is still possessed by people. They are important messages to consider if we as educators want to match the Activities of our pupils and our goals for them.

As the symbolic interactionist analyses the horizontal dimensions of human behaviour by going from one individual to another, the psychoanalyst studies the vertical dimension, trying to go beneath its surface in order to understand it.

Many of the projects I report in this book can be interpreted in the context of psychoanalysis and, I shall argue, the positive experiments we have made are due to the fact that we are pushed vertically by challenging oppressive forces. One example is the project of documenting responsibility (§1.3.). Children who could not say anything positive about themselves were helped to document, by means of tools from mathematics, the stronger sides of themselves: the various fields they, as members of a social setting, were responsible for. In a similar project they documented the principles of decisions. In which fields were they permitted to participate in discussions leading to decisions in various social settings?

The psychoanalytic dimension of such projects is not the documentation itself. Rather it is the discovery of the permission to discuss one's own strengths at all, followed by the use of mathematical tools for documentation. In general, projects which focus on oppression will include the psychoanalytic dimension. Projects about girls and their social role in the class, their choices of subjects, their chances of having a job and which job, etc., will all necessarily focus on the vertical dimension, including questions about oppression and the possible liberation from oppression.

Several linguistic researchers stress the decisive role of language as a means of power, social control, hindering or supporting Activity. Halliday (1980) emphasises how linguistic structure reflects social structure. He points out that, rather than saying that linguistic experience is the realisation of social structure, one should also take into account the converse relationship: language as a tool for such a structure, language as a metaphor for society. Language is thus a tool both for the transmission of society and for its transformation.

Wright Mills (1971, 1974) takes the same position. He argues that thinking influences language very little. On the contrary thinking has to borrow from social action which again is governed by forms of language. According to Mills (1974) no thinker can assign arbitrary meaning to his terms and be understood. Meaning is antecedently *given*; it is a collective "creation". In manipulating a set of socially given symbols, thought itself is manipulated. Symbols are imperative and impersonal determinants of thought because they manifest collective purposes and evaluations.

The problem is, of course, that the acceptance or rejection of collectively established words by others is a question about power. During the last decades we have had a huge body of sociolinguistic documentation about class differences and class hegemony in language use. One characteristic example of this is Bourdieu and Passeron (1977) who construct concepts like *symbolic violence* and *linguistic capital* in order to demonstrate how a ruling class can control an oppressed class by language use. We are going to examine such oppression closer, and relate it to Activity theory. First, however, back to basics: a review of Freud.

4.2.2. *Activity Theory and Psychoanalysis*

In order to realise the significance of Freud's theory we have to read him historically. This implies that we have to strip him of what was special to his time and what he met in his practice. I am thinking here, in particular, of the sexual frustrations of the Viennese bourgeoisie with which Freud was repeatedly confronted with via his patients. Moreover, I shall further disregard Freud's speculative concern about human drives as possible explanations for the distortions of sexuality which he observed.

By this approach we can discover the generality of the processes he describes and relate these to recent research in the field of human behaviour, especially linguistic behaviour. One might say that this leaves very little of Freud. Still, to him we owe the discovery of some extremely powerful conceptual constructs, from which we can profit as educationists. It is mainly the French structuralists who teach us to read Freud scientifically as demanded above.[2]

But even Mead, seven years younger than Freud, discovered the generality of his theory:

> Freud's conception of the psychological "censor" represents a partial recognition of this operation of social control in terms of self-criticism, a recognition, namely, of its operation with reference to sexual experience and conduct. But the same sort of censorship or criticism of himself by the individual is reflected also in all other aspects of social experience, behavior, and relations — a fact which follows naturally and inevitably from our social theory of the self.
>
> Mead 1934, p. 225

It is mainly Freud's concepts of the *conscious, subconscious*, and the *resistance* represented by the *censor* which are of significance for the vertical dimensions of Activity theory. Freud himself relates repression and resistance to forces of the Ego. The occurrence of resistance as the individual tries to prevent unpleasantness from becoming conscious, is well known to all psychoanalytic therapists. Apparently this resistance, represented by the censor guarding the gate between the conscious and the subconscious, is some kind of self-protective device inherent in the individual. However, had Freud examined his own writings as he describes the various clinical cases, he would, from his own words, have seen how the censor, as established in the individual, exists in terms of forces *outside* the individual.

Freud repeatedly uses terms such as "unpleasant", "indiscreet", "innocent", "meaningless", "embarrassing", terms which are clearly related to the judgements of the social environment. In Freud's own words:

> In all these cases there was a concern about a desire, which was in sharp contrast to the other desires of the individual, and which turned out to be contradictory to the ethical and aesthetic demands of his personality.
>
> Freud 1920, p. 26

I cannot but think that "ethical" and "aesthetic" demands have their origins in the social context of the individual, the context he relates to. We are thus back to the GOs. But with Freud we can take a step further than we could with Mead: the GO is also able to prohibit behaviour, to oppress behaviour and to favour and encourage behaviour.

Regarding the concept of *ideology* as a set of attitudes and rules taken over from the GOs, we can see the influence of ideology on repression: the repression can be considered as a result of present conflicts between the individual's knowledge about the world and the ruling ideology. According to this, the censor cannot be driven by forces within the individual. It is related to the GOs which have the possibility of functioning as an oppressive force. When the psychoanalyst feels a resistance against making experiences conscious, this resistance is due to claims made by the GOs rather than some built in "forces of the Ego". In this way we see the relationships between Activity theory, Freire's practical work and psychoanalysis.

Briefly the situation is as follows: Freire points to the lack of participation in Activities among the oppressed people of the Third World. He recognises the culture of silence as a result of oppression. He develops his method of *conscientisation* in order to provide his people with the means of becoming conscious of the oppressive forces, which, in its turn should lead to Activities combating them. Freire does not use the term "Activity", he uses the phrase *integration* in society as opposed to *adaptation* (see §1.2.3.). Activity theory offers an opportunity of which Freire does not take advantage. Although Freire repeatedly stresses that theory is of no use without being transformed into praxis, he does not take this to its full conclusion: he does not reflect on the next step for the people concerned, the transition from conscientisation to political action. Activity theory does so, as Activity theory includes man's transformation of his physical and social environment as an indispensable component. On the other hand, Activity can profit from Freire, as he clearly demonstrates how oppression denies man access to Activities.

4.2.3. *Ideological Forces as Repressive Forces*

Another relationship worth further examination in Freud's theory is the one between consciousness and the repressive forces. The paradigm is that if there is no repression, there will be no subconscious. Althusser (1970) declares the subconscious to be the object of psychoanalysis. He must do so, as *ideology* for Althusser is ever-present in society, exercised through the State Apparatus. For Althusser, ideology is a way of living, or *the* way of living. The role of ideology is one of cementation, as the legitimisation of the structure of society.

Ideology can thus only explain behaviour, not change it. A change of ideology will, according to this view, be due to political changes. The realisation of any political struggle is hardly possible in Althusser's theory. As the censors of the individual will be determined by the ideological forces, these forces are not regarded as basic: it is the resulting structure of the conscious which will be of interest for Althusser.

This points to the basic weakness of Althusser's theory (as with French structuralism in general): he offers a deep explanatory description of the relationships between the various components of society, connecting them so strongly with each other that one can hardly see how to turn his theory into any form of practice or Activity.

As educationists we have to look for openings for the adapted individuals, to use Freire's term. If distorted behaviour, that is behaviour which is due to the repression of conflicts between ideological forces and the experiences of the individual, we have to challenge this repression. In order to do so, we must focus on the repressive forces, rather than on the behaviour which results from these forces. Such a critique applies equally to Wright Mills, who describes language precisely as a form of social control, but at the same time describes it as a somewhat autonomous system where the controlling powers stay controlling, without considering the possibility of challenging them.

Both Freud and W. Reich assert the presence of repressive forces. Freud does it by his psychoanalytic praxis (attacking ideologies about sexual behaviour), Reich by his politicisation of the subconscious:

> The censor is not something mystical. It contains rules and prohibitions which are adopted from the outer world, and itself becomes subconscious to the individual.
>
> Reich 1969, p. 21

The repression is related to this subconscious, and the goal of psychoanalysis becomes just as much to uncover this unconscious process as to uncover what is kept hidden by it. Reich expresses the task of psychoanalysis as the study of a dialectic movement: repressive force —

repressed knowledge — repressive force . . . , a progression through the abolition of a pattern of contradictions. So both Freud and Reich were perfectly aware of the importance of discovering the repressive forces, or, in our terminology, the forces which deny access to Activities.

4.2.4. *Repressed Knowledge Does Not Disappear*

One of the most significant findings of Freud which proves extremely important for education is the phenomenon that repressed knowledge does not disappear. This observation can be made by anyone sensitive to those nearest him, and is the killing argument against behaviour modification theory: you treat one symptom by modifying behaviour, and beyond the reach of the therapist another symptom arises.

Freud explains that repressed experience remains in the unconscious and signals itself to the outer world in various ways, leading to behaviour which may be noted by an outsider as distorted. It is this position which leads Lacan to state that "the amnesia of repression is one of the most lively forms of memory" (Wilden 1968, p. 23). In other words, the individual remembers through behaviour. It is here that the major contribution of Reich to psychoanalysis is centred: the role of the body in repression, as the location of various knots in the individual, knots tied as the result of repressive forces. The body remembers.

A repetitive history of repressive forces functions as sedimentation in the individual: as the repressive forces get the chance to work over time, again and again they bury the prohibited experiences deeper. If this history is short, the repressed knowledge can be just under the skin; easy to uncover in a suitable context. On the other hand there can be repressed knowledge which is congealed so deep down that it will require correspondingly strong forces to help the individual to reach them.

It is the existence of such repressed knowledge that Freire takes advantage of when he provides the silent culture of the masses with an educational context in which they can speak. Freire knows about the character of oppression and the resulting silence. He lives and works *among* those who are to become his pupils. Thus he has some joint experiences, some insight into the daily struggle to cope with reality. He thus concentrates his efforts precisely on the vital field of knowledge for his pupils: the knowledge they possess, but which they are denied by the authorities.

Freire's method is an example of the significance of the methods of social anthropology for education, especially the method of participating observation: in order to know about which educational tasks can tie up with the Activities of the individuals, it is profitable to observe a culture from the inside rather than the outside.

The view that repressed knowledge does not disappear leads some theorists to argue that the subconscious does not exist. Volosinov (1976) reduces psychoanalysis to a problem of ideological hegemony:

> The whole of Freud's psychical "dynamics" is given in the ideological illumination of consciousness. Consequently it is not a dynamics of psychical forces but only a dynamics of various motives of consciousness.
>
> Ibid., p. 77

What is said and not thought is not held subconsciously. It exists: it does not just come to the surface because of ideological suppression, according to this view. In our context here it seems that the issue whether the subconscious exists or not is purely an academic one. The important thing for the educator is that what the individual or a group of individuals does not say directly can exist, possibly just under the skin, ready to be unveiled. It may furthermore be just this phenomenon — that what it is important to talk about is not said — is the reason why we see no Activity.

If man is prevented from engaging in Activity in this way by continous and strenuous oppression, he does not necessarily die, in the physical sense of the word. He just fades away psychologically, becoming a member of what Freire calls the culture of silence.

4.2.5. *The Unconscious as a Language*

To summarise so far: the existence of oppressive forces is significant for man's possibility of participating in Activities. Linguistics contains promising theories for the analysis of such oppression. Mordern psychoanalysis, as in the tradition of structuralism (Lacan), works according to the thesis that the subconscious is structured as a language.

As I have already made clear, I do not see the problem of the subconscious as important. The importance for Activity theory lies in the phenomenon that the ideological superstructure of language suppresses man's possibility of realising himself through his Activities. As Volosinov stresses:

> Those areas of behavioral ideology that correspond to Freud's official, "censored" conscious express the most steadfast and the governing factors of class consciousness. They lie close to the formulated, fully fledged ideology of the class in question, its law, its morality, its world outlook. On these levels of behavioral ideology, inner speech comes easily to order and freely turns into outward speech, or in any case, has no fear of becoming outward speech.
>
> Volosinov, op. cit., p. 89

I hope my reader sees the crucial relevance for education: following Volosinov here, the strategy for those pupils who have ceased to learn the subject of mathematics (or other subjects) will be to concentrate on subject matters which are related precisely to the inner speech which is not permitted to become outward.

Sociolinguistics provides much empirical research which can be interpreted in the context of psychoanalysis. The person who, without doubt, has carried out the most thorough research in the field over the last decades is the American William Labov (1972, 1972a, 1972b). He points to the need to reorientate current segmentations of the study of man in order to observe what so far has been pronounced as unobservable (Labov 1972). Thus, regularity can eventually be obtained where previously confusion reigned.

Labov continually stresses the *self* as the root of linguistic change: self-consciousness is the basic factor related to linguistic behaviour. Labov (ibid.) found people unaware of their own language use except for the stigmatisation by the larger community. Accordingly he sees the participation in linguistic change as a sensitive index of social mobility and struggle. Linguistic change therefore has to be examined in terms of society, as well as of the psychology of the individual.

We observe the importance of the relationships between linguistic change and the politics of the environment in the case of women's use of language: females discover how their use of language is freed from oppressive forces as they fight their way towards womens' lib. In the campaign for womens' rights they simultaneously change their linguistic behaviour: at public meetings they debate where previously they would have kept quiet, they start questioning at college lectures where previously their male fellows were the best at posing problems and so forth.

Rosen (1972) makes parallel observations in the case of workers. In a critical look at Bernstein's theories, he points out that

> the most articulate workers are those who have actively participated in the creation and

maintenance of their own organizations, and amongst those the most articulate would be those who in that process had encountered and helped to formulate theories about society and how to change it.

Ibid., p. 9

The relationship between use of language and participation in Activities is obvious from this quotation.

4.2.6. *Linguistic Registers as Oppressive Forces*

Halliday (1978, 1980) provides us with a useful tool for our purpose. That is the concept of a language *register* as opposed to a *dialect*. A *register* is the way to say things within a specific field. Edwards (1976), for instance, examines classroom registers. Registers thus refer to language use which is situational: we can master (or not master) situations by our use of speech in the classroom, in the gang at the street corner, at a cocktail party and so forth.[3]

Registers are related to *use* as dialect is related to the *user* of the language. Dialect is the form, different dialects can express the same content in different ways.

Halliday (1978) describes a register as a set of meanings that is appropriate to a particular function of language, together with the words and structures which express these meanings. To develop language then means to add to its range of social functions: developing new registers. A dialect is a means of expressing class-consciousness and political awareness. Halliday (1980) uses as an example a society split into two conflicting groups: society and antisociety and — correspondingly — standard and non-standard dialect. In a hierarchical society dialect merely becomes the means by which a member gains, or is denied, access to certain registers.

It does not appear from Halliday's writings whether he considers dialect and register to be two aspects of the same case: *ideology*. Registers *and* dialect are vital for both communication and social control. The classroom is an example of this: Pupils usually have to talk properly in order to be permitted to take part in the daily tasks. Dialect thus functions as the form of communication in the Activity. They would reflect the political location of the Activity: social class, rural or urban, subculture A or B etc. Registers, on the other hand, would represent the political content of the Activity.

It is hard to imagine otherwise than that form and content relate to each other complementarily: what it is legal to say and in what way it has to be expressed. Form exists in the context of content and conversely.

In his appraisal of Bernstein's work Halliday (1978) stresses how Bernstein's findings about the differences among young people in London were mainly related to registers, not dialect. He points to Bernstein's greatest achievement: relating social classes, family roles, registers (as semiotic codes) and orders of meanings and relevance. The point is well taken, as it tells us that if we want to design educational situations inviting Activities, it is not enough to listen to how pupils speak, but also what they speak.

Halliday (ibid.) also hints indirectly at a major problem which I am moving steadily towards and which I can hardly do more than identify: the existence of mathematical registers. Such registers must not be confused with mathematics itself (although Halliday calls them the language of mathematics). They comprise the kind of intermediate language, placed within the spoken language, which serve as the starting points for mathematisation. They are the linguistic forms which represent some problems and which are the final linguistic version before mathematisation. In short: the "mathematics" in the language of an individual constitutes the mathematical registers of that individual.

Just as Folk mathematics (§1.1.) builds the *material* with which mathematisations operate, mathematical registers do the same. When Thomas (7) finds it worth telling us that he will always be 2 years older than Lars (5), this can be the foundation for both $7 = 5 + 2$ and $y = x + 2$.

The ideological and possibly oppressive content of mathematical registers is difficult to observe. It is far easier to identify how such registers relate to culture. I hope this has been demonstrated in Chapter 3: different groups of people speak differently about their world, and accordingly they develop different material appropriate for mathematisation and mathematical modelling.

The notion of register raises difficult problems: what is the nature of the relationships usually modelled in mathematics as seen from the perspective of registers? What is the ideological content of the relationships which are usually modelled in mathematics education? What is the ideological colour of the use of thinking-tools as used in the wide sense? Halliday (ibid.) tells us that every language possesses the potential for developing registers for mathematics — and science and technology. Pinxten has demonstrated this in the case of the Navajos. But what about those kids from the backstreets of large cities who definitely code the vital experiences of their lives differently from those designing the mathematics curriculum? What would the implications be if we really made an attempt to make a bridge between their most commonly used registers and those of mathematics? What would the problems look like then? What about the projects? And what about the various mathematisations — would they represent the mathematical structures commonly taught today? If Sinclair (1983) is right when she stresses that failure to learn in school is usually based on problems of language rather than conceptual deficits of the child, the *register* can prove a useful tool for the educationist.

Halliday, Labov and Rosen all, in their own way, describe and interpret the social function of language. They furthermore discover the phenomenon that when language use is oppressed, language does not necessarily disappear. Ghetto languages, languages of protest and resistance emerge. In psychoanalysis we can consider such languages as signs of resistance: oppression leading to repression does not imply that the experiences of the individual disappear, they remain in more or less distorted forms. In the cases which Halliday and Labov describe, the language users possess knowledge about the oppressive forces, leaving them in a much better position for Activities than in the case of Freire's illiterates.

Mathematics is also less favoured than language as "protest mathematics" hardly exist. There is Folk Mathematics which is something different. Folk Mathematics exists outside school mathematics and is used in the daily Activities of the individual. It is not developed as a means of protest and resistance in order to cope with domination and oppression. According to the corollaries of the theory of the I- and S-rationales, the pupil of mathematics who decides to stop learning the subject has not been in a position to develop other alternative uses of the subject.

4.2.7. *Summary*

Freud's contribution so far seems to be the discovery of some important aspects related to the influence of society on the individual. Being a psychiatrist, he would mainly come up against cases of pathological behaviour. He built a dialectical theory, as he also looked for explanations outside the individual as well as inside. Furthermore he discovered the phenomenon that will damage education until it is accepted and assimilated into educational theory and practice: knowledge repressed by oppressive forces in the individual's environment does not disappear. It still exists, eventually showing itself in a distorted form.

What do we find if we return to the mathematics classroom, bringing these research tools with us? It depends, of course, where we are. In a Western classroom we will usually see busy pupils working at tasks, whatever they may be, as long as they are provided by their teacher. These pupils obviously have no difficulty in relating the tasks to their Activities. We can also see those who have stopped working on the subject, looking into the air or drawing flowers or machine-guns until their teacher notices.

We see those who stubbornly keep trying to get the hang of it and usually get it all wrong. Sometimes they produce methods and solutions apparently out of the blue, still pretending that they are participating in some game which is of the utmost importance to them. Finally, we have a group of pupils who are frequently recognised as the real problem of the class. It is those pupils who mess about, making the most of it. Ultimately they might be removed from the class, perhaps as cases for the educational psychologist.

We hear about low attainers, low achievers, children of low mathematical ability and slow learners. Looking closer at such labels, we usually find that they are given to pupils who have not been able to keep to certain *standards* set by some authority or other and which are to be reached within a given set of frame-factors.

The pupils who dropped out of the examination course have often not seen the subject as providing sufficiently important tools for their past, present and future. Or they may once have done so, when they started school, and at some time stopped. I offer Activity theory as a general approach for tackling such situations where both passive and rebelling pupils can be identified.

Psychoanalysis takes into account the biography of a person and the effect on that person of the oppressive powers. Contrary to Activity theory, this is an individualistic theory. The most important difference, however, is that, contrary to psychoanalysis, Activity theory is oriented towards the future. It stresses the future life and possibilities of individuals, helping them arrive at a position where they can be constructive about their own life-situation. While the psychoanalyst follows his client to the threshold of his clinic, offering him the promising prospect of future life, Activity theory is *in* the future life. Whereas the psychoanalyst helps the individual to identify the powers that oppress him, Activity theory implies a direct attack on these powers.

Psychoanalytic research has demonstrated the effects of oppression. Although the various case studies reported can look hopeless and tragic, they leave one optimistic and vital possibility for the educator: oppression leading to repression only buries the potential for some potent learning. It does not eliminate it. It is this repressed potential in the form of repressed experience taken together with the repressive forces which are a key to Activities. Or to the design of educational tasks which will connect with Activities.

4.3. COMMUNICATION THEORY

4.3.1. *Institutions Communicate*

> "What kind of kids are they who run away?"
> "Vigorous natures who don't thrive."
> "But if they don't get the opportunity to escape?"
> "Well — they apparently become like most people, just a bit more damaged."
> The captain and the mate in Jens Bjørneboe's *Jonas*

The English social anthropologist Gregory Bateson contributes to education a strong and versatile communication theory.[1] In what follows, I may be pushing his theory somewhat beyond his conceptions in my consideration of the structural properties of the communication of institutions, including the double-bind. Not only individuals communicate, institutions do as well. Bateson only rarely, if ever, considered the field of education. But he spent sixteen years of his life occupied with psychiatry, building his theory of schizophrenia on the contradictions in communication and learning.[2]

The educator like myself, who is concerned about the rebellious pupil and the silent pupil in a closed communication system such as a school, seems to have some promising thinking-tools in prospect. And certainly the tools turn out to have the power to understand and cure much of the pollution produced within such a system.

An institution such as a hospital or a school can, in the same way as a GO, be regarded as a sender of messages to the individual. Just as the individual relates to a system of GOs he is also connected to a communicative network. It is the various messages within such a network that Bateson and his followers teach us to interpret.

4.3.2. *Metalearning*

Bateson distinguishes between learning and metalearning (Learning 1 (L1) and Learning 2 (L2)) and the relationships between these. Metalearning is learning about context. Context can be just everything which embodies knowledge in some learning situation. I cannot go into the detail of an analysis of important contexts for learning without finding myself writing another book. To give the context is to provide examples to illustrate a concept. It is the theoretical context we build in order to structure the appropriate thinking-tools. Fractions can be embodied in the context of rational numbers, vector calculus in an algebraic context.

The teaching of a subject can be practised in the context of a specific didactical theory calling for inductive learning, a problem-solving approach or a learning-by-discovery approach. The teaching of a subject also depends on various frame factors which build context at another level: class size, number of lessons compared with the length of the syllabus, etc. Various social contexts can be identified and I shall give examples of their significance for learning throughout this section. So there certainly exists a network of contexts. It is an important field for research (and for student-teacher projects) to explore the hierarchical structure within such networks. One problem which is clear from the above list is that there do not seem to be any limits to how wide a context one can examine: the contexts of curriculum, classroom, school, town, nation, . . . The crucial problem is to locate the boundaries of the educator's control: which contexts are within our reach as teachers and which are not?

When facing a learning situation, the pupil encounters a set of *context markers* with which he is more or less familiar. In other words, the pupil has some previous experience from other situations where he has faced some of the same context markers. These experiences constitute his *metalearning* for the situation. He applies his metalearning when facing the familiar context markers in the new learning situation. This metalearning will be decisive for the way he will relate to that situation. Metalearning is thus learning about context.

After metalearning we obtain *metaknowledge*, that is, knowledge about knowledge.

To give some flavour of the significance of metalearning for education I will report one of Bateson's own expositions. But first, a personal experience.

At that time I was a regular visitor at the research seminars of the Institute of Cognitive Psychology. This had specialised in problem-solving, and as I was a fresh teacher in the comprehensive school, it should have been very rewarding for both the Institute and for me. And indeed it was. I could help them with the ever-recurring problem of the social scientist who works experimentally: I could provide the subjects for their experiments.

My fellow teachers were more than pleased to give their classes to me for a session or two for what I presented to them to be of utmost significance for future education.

The research usually consisted of some pencil-and-paper test about the hat-rack problem. For some reason 55% of American College Students at that time (the mid-sixties) had solved this problem and it was interesting to find out the characteristics of their cognitive skills.

In short, the problem consists of providing equipment on which you can hang a hat. You are in a bare room, and have at your disposal two sticks, say each 1.5 meters long, and a vice with which to join them together. Sometimes the problem was presented in a realistic setting with the equipment present. Usually it was a pencil-and-paper test (a) with a drawing of the equipment or (b) without a drawing. I shall not give the solution. I would rather talk about the faces of the pupils.

"Oh no, Teach. Not another test. You gave us that extra vocabulary yesterday."

"Teach, we have just had gym. We are exhausted."

"The sun is shining, and you bring us a test. What are you trying to do to us?"

"Who is that little sweet fellow? Is he a new one? Oh — we will take care of him. You just leave, Teach."

"There is Mellin as well. What the hell are you going to do with us? Last time we spent two hours doing your funny things."

"There are three of them. We must have quite a reputation now."

Their teacher withdrew to the promised cup of coffee. The graduate student or the research assistant remained with the class, and the test papers were distributed.

"It is all anonymous and will have no effect on your marks."

I walked around, saw someone really trying, some flowers and some destroyers being attacked by Fockers and Messerschmidts. The first aircraft, made of a folded test paper, was let fly.

My last memory of one of the classes is of a research student running around the class at the end of the session, collecting papers, unfolding them and putting them into an envelope. The data were ready for coding.

At that time some of us read American ethnomethodology with hopeful eyes. One of the major messages here was that the positivist's application of the methods of the natural sciences was bound to fail as the scientist in this tradition made attempts to eliminate context factors and as it was precisely these which were of significance in order to understand human behaviour. Then Bateson (1973) arrived, stressing exactly the same but much more consistently. And he brought new hope to frustrated educationists.

Bateson (ibid.) analyses an animal experiment in the Pavlovian tradition. It consists of a series of learning situations where the animal is to distinguish between an ellipse and a circle. When such discrimination has been learned the experimenter changes the shapes of the figures: the ellipse is made progressively fatter and the circle is flattened. After a while a stage is reached where further discrimination becomes impossible. At this stage, the animal starts to show symptoms of severe disturbance. Pavlov related these symptoms to context and to metalearning. The context, here the laboratory (smell, noises, furniture, . . .), and the type of learning situation (discrimination), develop a metalearning: in this laboratory one has to discriminate between figures.

Then there is a change in context: the content of the situation does not require discrimination as this has become impossible. Rather it calls for some kind of guesswork, which according to its metalearning the animal is not prepared to perform. The next experience is that the "experimental neurosis", as Bateson calls it, disappears in the nearby park, even if the animal is confronted with the same experimental situation there. The resulting thesis is that by changing the context, cognitive, social, environmental, etc., we change the learning situation.

4.3.3. *The Dialectics between Learning and Metalearning*

The major thesis of our theory so far is that there always exists some metalearning which the individual applies when facing some learning situation. No learning exists without some metalearning. The pupil may be right or wrong according to the expectations of his teacher, but the scent of certain context markers whenever he faces a learning situation turns on certain strategies. His former experiences accompany him as he participates in the learning (or not).

It is amazing to see how this simple principle is neglected in mathematics education. We see it particularly in those cases where the textbook or the teacher shifts context without letting the pupils know. A familiar example of this is in the case of fractional numbers, where addition and subtraction goes easily as the pupils operate on fractions of some unit, and multiplication is not understood, as this requires a shift of context such as to the field of rational numbers. Addition originally made things bigger until we met the integers. Similarly, the product of two rational numbers can be smaller than the numbers themselves.

The principle of discussing the context with the pupils, what we call discussions at the metalevel, is decisive for how the pupils relate to the knowledge they face. The argument is, as stated above, that they will always relate knowledge to some context, and that we would thus be better off helping them with this than neglecting it.

Returning to the discussion about the S- and the I-rationale in §4.1., I can now reformulate the hypotheses made there, saying that the pupil, when facing some learning situation, recalls some metaknowledge which determines his attitudes to the situation. In the main there exist two extreme forms of metaknowledge: mathematics as something which is experienced as a set of tools for Activities (S-rationale), excluding School as a possible Activity, and mathematics as something which belongs to the context of school and only that (I-rationale).

The theory developed so far may lead to determinism, as a certain metalearning can be interpreted as being decisive for, and restrictive of further learning.

Sometimes a struggling teacher can also experience the metalearning of his pupils as something which cements their learning behaviour, out of reach of his wishes to develop new attitudes, new rationales or new conceptions about the subject. Clearly it is not like this.[3]

There exists a dialectic between learning and metalearning which we have to consider. I shall explore this by giving some examples. Take a teacher in a secondary school. He faces a new class. He has some fresh ideas about what he wants to do, and these may conflict with the kind of teaching the class is used to. We then face the following situation:

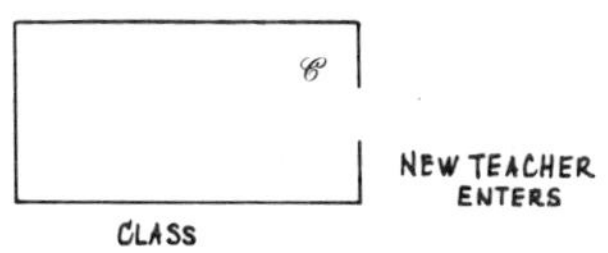

Fig. 4.3.1.

The 𝒞 here represents not only the class, but also the metalearning existing in the class about mathematics education (or education in general including mathematics.)

The teacher carries his ideas, which if he starts them working immediately, will expect a different metalearning in the class:

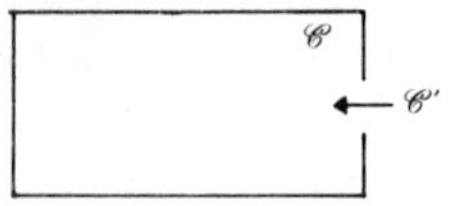

Fig. 4.3.2.

In this case he will probably face some difficulties:

"Teach, we've never done mathematics like this."

Rather than introducing a completely new set of context factors simultaneously, the teacher can build niches, each representing new sets of ideas about mathematics at the metalevel:

Fig. 4.3.3.

There is a must in this strategy: as the pupils continually develop their metalearning the teacher has to communicate his ideas as represented by $\mathscr{C}'$. The concept of a double-bind is operating in the shadows here, because this concept relates to whether the individual has the appropriate metaknowledge or not. By extending the niches, thus developing $\mathscr{C}'$ at the cost of $\mathscr{C}$, the teacher can gradually transform the metalearning into the one wanted.

There is another possibility. That is to break the double-bind, crash it, and start afresh in a completely different context. In situations where hard resistance against further learning occurs this may be the only possible solution.

I use *resistance* in connection with oppression. In order to crash a double-bind, or to transform resistance into cooperation, the educator must thus consider the oppression, so as not to reproduce it.

I have already reported on experiences from vocational school where the pupils suddenly realised that they had been cheated. They did mathematics on the shop-floor, and then moved into the classroom, thinking that it was theory as it said on the timetable, rather than mathematics.

When I myself face school rejectors I usually take them out of the classroom in order to change as many context markers as possible: remember Tommy and Tage Werner's experience recounted in the Introduction.

The last group I had was what we then called an "observation class" in Norway. Nowadays such a group will get a less stigmatising label (so it is claimed), such as "clinic", "resource group", or "social group." I brought them out of the classroom to study illegal parking in the narrow streets around the school. The dangers such parking caused for pupils going to school were severe. We brought the statistics and the geometry of it into the classroom, and pursued the mathematical side of the project there. Finally we produced a neat report for the local newspaper.

We went on. There were five boys altogether, searching the district for empty flats. We became quite experts at identifying them, seeing how the landlords made attempts to camouflage them. (Plastic flowers, and did you ever see two flats with identical sets of curtain?) Again we brought our findings into the classroom, where we analysed the distribution and density of the flats. We did not report these statistics to the newspapers, but the work gave rise to some interesting discussions about housing conditions and political matters connected with them. Not surprisingly, the boys had better insight into this than I had.

The degree of success of my innovative work was revealed by the question made by one of the boys one sunny day in the spring:

"What are we going to count today. Stieg? I think we should all go out and count the nicest lassies."

The principle exemplified here is to build a completely new learning context located outside the classroom. This accords with the hypothesis that for some groups of rejecting pupils, the classroom provides such negative contexts for learning that there is little chance of initiating learning. It is even a question of whether school in general provides negative metalearning which prevents participation.

Examining a suitable context for knowledge outside the classroom, examining the possibilities of thinking-tools inside the classroom provides an opportunity to connect outside mathematics with inside mathematics. In the case of the vocational school this worked even better, as the pupils there could also explore the thinking-tools on the shop-floor: the pupils in the welding class could construct the angles and the parallels on the steel plates.

4.3.4. *Children's Metaconcept of Mathematics*

Pupils have a metaconcept of mathematics and of mathematics learning. Evidence about such concepts can be experienced from pupils' statements about mathematics, which are usually statements at the meta-level. Pupils usually speak of their general experiences when questioned about mathematics, and if they mention one particular lesson, it is often in contrast to what the lessons are usually like.

The lad in the Introduction told his teacher, in this case the author, that for once he was grateful for the lesson as it gave him new faith in school.

Hoyles' (1981, 1982) works are packed with evidence of pupils having metaconcepts of mathematics and mathematics learning:

P1: Well, there is maths. I always find Maths hard. That is why I switched from O-level to CSE because I found the O-level too hard. Maths is my weakest subject and I'm useless at it.
P2: Well, I just seemed to be able to do these triangles. It was amazing because I'm usually no good at maths and way behind. Every question came along and I just did it O.K. It's not like that now, I can't do anything and find it awful.
P3: Oh yes. In maths, well we did these cards you know. That is the only thing I could do, so you know, I liked doing them. I could never do it before.

Hoyles 1981, pp. 5—8

My point is that children relate to mathematics as a totality and not as a discrete set of events. It is such a totality the educator eventually has to challenge in order to change the state of affairs.

Some major types of metaknowledge about mathematics and mathematics learning can be grouped under headings like:

- I am (not) a person who can learn mathematics;
- I do (not) usually learn mathematics in this kind of situations;
- Mathematics is (not) (a) important as a field of knowledge that I can use outside school (b) important as a means for the future I hope for.

In §1.3.6. I described the attitudes that may be evoked by someone receiving a string of tools $x_1, \ldots, x_n$ which he is told have some usefulness not immediately apparent. The point was that an individual will accept some tools of this sort and then he will resist any new ones. The individual's evaluation of usefulness, and the length of the string he is thus willing to receive, is dependent on his metaknowledge about it: to what extent does he recognise the possibility of it becoming useful?

The *functionality* of one of the tools may imply the functionality of the complete string. That is: the individual's conception of functionality of the various elements in the string of tools relates to his metaconcept of the string. I am now using *functionality* in the sense I analysed it in §1.2. There is thus a dialectic between the length of the string of tools a pupil is willing to receive and the recognition of the usefulness of the particular tools as they are presented to him.

The metaphor used here is not adequate, as mathematics is much more than strings of tools. The metaknowledge about a string of mathematical tools must include knowledge about its structural properties. To have a functional metaconcept about mathematics, that is to appreciate mathematics as functional knowledge, implies more than the appreciation of the mathematical tools as functional tools. It implies as well an appreciation of the importance of the way they relate mathematically to each other by the logic and reasoning specific to mathematics. In the end it is this which builds the various tools and makes them possible.

4.3.5. *The Double-bind*

The next exciting feature in Bateson's theory is his conception of the double-bind. By this concept he generalises Freudian theory into a general communication theory:

> Freudian psychology expanded the concept of mind inwards to include the whole communication system within the body — the autonomic, the habitual and vast range of unconscious process. What I am saying expands mind outwards. And both of these changes reduce the scope of the unconscious self. A certain humility becomes appropriate, tempered by the dignity or joy of being part of something much bigger. A part — if you will — of God.
>
> Bateson, op. cit., p. 436

The importance of the double-bind for education, as I have stressed several times before, is that it is a concept which helps us to examine the *socialisation* process, as socialisation can always be considered as a double-bind. I shall introduce the concept of *level* of the double-bind, indicating that if this level is set too low, the individual will be tied to conflicting messages, become confused, perhaps paralysed in relation to Activity.

Bateson's concept of the double-bind is not at all clear or consistent as he develops it over time. His first discovery of communication was when he studied mother/child relationships in rearing situations in Bali. That was as early as the thirties. His first book *Naven* (1936) was probably 30 years ahead of its time, and according to Bateson's reviewer Kuper (1983), achieved no more than an underground reputation in British social anthropology.

Bateson's discovery was that the Bali mother applied a teasing-attracting pattern of behaviour towards the child as a method of upbringing. It is such a method of shifting the mode of communication which creates the double-bind, the child being bound by contradictory messages, positive and negative reactions to the same sort of behaviour.

What Bateson learned later was that this sort of socialising process created schizophrenics in Western societies while it produced harmonic members of Bali society. The original double-bind situation is thus the mother-child situation where the mother simultaneously sends positive and negative messages. The classic example is the situation where she stretches her arms saying some nice things to the child while at the same time her facial gestures communicate rejection. As we shall soon see, this original conception of the double-bind unfairly stigmatised the role of the mother as a possible "cold and rejecting" mother with a damaging influence on the child's psyche.

Sometimes Bateson uses the concepts of *analogue* and *digital* communication in this context. A digital communication consists of signals which are discrete (speech consists of a discrete set of sounds), an analogue communication consists of signals that are continuously distributed (like the volume of the voice). Instruments such as the piano and the viola may serve as examples, communicating respectively by means of discrete and analogue sets of signals.

A double-bind can thus occur as a result of a contradiction between a discrete set and an analogue mode of communication from the same sender: the discrete sets of words taken together with the warmth of the voice, for example.

In general, the double-bind builds on the presence of contradictory communications. Furthermore, as it is a *bind*, the receiver of the communication is not aware of the contradictions involved, he has not the appropriate metaknowledge about the contradictions present. Therefore the confusion and paralysis. We should also note Bateson's stress on repetition as a condition for the double-bind to be damaging. This is connected to the same principle in psychoanalysis: it is not the presence of the oppressive forces which causes the problems, it is their repetitive use which may be fatal.

I shall now analyse the sending of contradictory messages in more detail. It is not only the mother who communicates, the father does so too.

What about the joint communication from the parents: does mother say A when father says B? What about the communication between mother and father and the communication this represents to the child? Is it possible for a child to relate rationally to a mother-father relationship which is imposed by conflict without having metaknowledge about that? To what extent does the child not only relate to its mother (father) but also does this through the other parent? Continuing in this way, we can develop a matrix of communication and examine this matrix for contradictions (Ruesch and Bateson, 1951).

Taking a step further we can look at institutions in general. What about the communication from the doctor of a hospital as compared with the communication of the nurses? What about the communication from teacher *x* in a school as compared with the communication from teacher *y*? In order to analyse communication dialectically, which Bateson does not do sufficiently in his writings, in my view, I shall also have to examine the material frames of an institution, and school in particular, to see how this produces certain ways of communication.

4.3.6. *Double-binds in Education*

The French mathematical educator Guy Brousseau has constructed a didactical structure based on a double-bind: the didactical contract (Brousseau 1984). In its simple form the contract means that the teacher is obliged to teach and the pupil is obliged to learn. For the persons involved, the contract cannot be negotiated. Teacher and pupil are thus all the time busy inventing ever new forms of behaviour and interaction, which they hope can be in accordance with the contract, which are either interpretations of it or tolerable evasions.

Such a contract is a double-bind, as pupil and teacher are locked in an expectation of growth which can be established by circumstances outside their control. Structurally, there would be no difference if the contract stated that the pupil was to grow 5 cm the next year. The brilliance of Brousseau's structure is, however, that in order for genuine learning to happen, the contract has to be broken. In that case teacher and pupil manage to crash the double-bind:

> To be obeyed, the contract must be broken, because knowledge cannot be transmitted ready-made and hence nobody — neither the teacher nor the pupil — can be really in command, can really control the contract. (Brousseau and Otte 1985).

Brousseau is courageous in basing his didactical theory on a double-bind like this. According to the view that the real progress of humanity can be considered as a series of crashed or defeated double-binds, he is right in basing his key concept, the contract, on such a structure. What makes his conceptual construct so fragile is that the didactical situation does not consist solely of the teacher, the pupil and the subject matter. The didactical situation is usually a power situation; and people and institutions outside the classroom also exercise power. The didactical situation thus relate to a larger structure. If the power held by factors outside the classroom can have an influence inside the classroom, the didactical contract can similarly be influenced by factors outside the classroom. In such a case it is hard to see how the problem of breaking the contract can be solved in the context inside the classroom.

An institution does not only communicate by the private communication of its members. It also communicates according to its frame-factors, regulations and instructions. The frame-factors cause the members of the institution to communicate in specific ways. The study of this is one of the major objects of ethnomethodology (Douglas 1973). The aim is to see how the regulations of an institution create routines and common sense knowledge within it. My project is to examine the production of double-binds as the result of such routines. Such a double-bind rules among most of the members of an institution. In school a teacher hears from one authority that he should develop attitudes of a certain kind towards his pupils while at the same time he receives messages from other authorities which clearly contradict the former. I have often wondered to what extent my advice and convictions on mathematical education as I have formulated them for student-teachers and in my writings have contradicted the other messages they receive: examinations; get through the book during the year; get decent results in the standardised tests and so forth.

The most prevalent structural double-bind in Western schools is probably caused by the contradiction between examination systems calling for individualistic strategies and the ideology calling for co-operative methods.

There will be further variations in the messages sent from this structure according to whether there is (a) a job or a place in the continuation school for those who pass the exam or not and (b) an examination system letting 50% or 95% of the applicants pass. Accordingly I regard the marking of a test as a communication and the brief comments of the teacher as he returns a test similarly.

There are lots of variations on this theme. The point has been made: there may be important structural properties within an institution which promote certain forms of communication, which in the end may cause various double-binds.

The lack of concern by Bateson to consider how the frame-factors of an institution can constitute a specific structure of communication also implies a lack of concern about the significance of the material factors for communication. One of the few times he considers these is in the case of the Bali ethos:

> It is immediately clear to any visitor to Bali that the driving force for cultural activity is *not* either acquisitiveness or crude physiological need. The Balinese, especially in the plains, are not hungry or poverty-stricken. They are wasteful of good, and a very considerable part of their activity goes into entirely non-productive activities of an artistic or ritual nature in which food and wealth are lavishly expended. Essentially we are dealing with an economy of plenty rather than an economy of scarcity.
>
> Bateson, 1973, p. 89

The double-bind functioned positively as a means for socialisation in Bali society. It does not do so in Western society. I shall relate this to the *level* at which the double-bind is exercised by those having the power to do so. Socialisation implies the adaptation to the norms of society at the same time as the individual has to cope with the demands of that society. Socialisation thus contains a contradiction between control (norms) and individuality in the setting of various frame-factors (as an economy of plenty and one of scarcity).

Such control may be exercised on too low a level preventing the individual from coping with the demands of the outer world. A typical example of this is the family widow. This is a category defined within social medicine. A family widow is a person who was once chosen by the parents to stay, to function as a maid for their old age. When the parents finally die, the widow will often be in her early fifties. She is then often unable to fend for herself and is usually offered a pension.

Control may also be exercised on too high a level. In the case of a pupil's private arithmetic (§2.3.) it is not very useful if each pupil has his own arithmetic so that the communicative aspect of arithmetic does not operate.

The determination of the proper level for a double-bind is a subtle and difficult task for any educator. The method of avoiding its damaging effects is to loosen it by communicating at the metalevel as often as possible, thus releasing the contradictions which determine it.

4.3.7. *Responsibility*

Responsibility is a key concept for educational Activity. "It is all up to you, but you have to be responsible for your actions" is what sons and daughters of every family have heard over the centuries. And from their teachers as well. The statement contains a double-bind. The problem is whether we, as controlling adults and educators, offer too much responsibility or not. We want Activities, and this implies that the pupils have to do something on their own, creating their own projects and being responsible for their organisation. "Strong" or "independent" individuals relate to the demands of the social environment, i.e. society, and the individuals should thus get some opportunity of testing themselves against these demands. My belief is that we as adults are often too reluctant to give away responsibility for children to have such experiences. Education has few traditions of providing pupils with responsibility in the full meaning of the concept. There is one particular field in which it has become important for education to demonstrate willingness to delegate responsibility. That is in the field of drug education.

When young people face drug situations they will be on their own. There will be no parents and no teachers to protect them. They will have to be strong enough to cope with the situation, stand up for themselves and reject the various offerings. The only thing the adult world can do is to provide the youngsters with sufficient thinking-tools and communicative tools, together with sufficient responsibility to make them secure in the drug situation.

One method of activating young people in their own fight against drugs in Scandinavia is the "Beat the hashish" campaign. This was originally directed by the youngsters themselves, and it was approved of by the professional authorities in the field. It was a clean-up campaign, built on the use of badges, discussions, demonstrations, Rock against drug, sober discos and so forth. At some time they were also directed towards campaigns for better social environments, including the demand for more jobs and more youth clubs.

I shall return to the role of mathematics in such campaigns in §5.2., but we can perhaps already see how the subject becomes important for surveys, budgetting, mobilisation strategies etc.

The campaigns so far have been quite successful in several places. The significant point to note, however, is the appointment of consultants for such campaigns by the State. If these are to function in a context of control — and it is difficult to interpret this initiative otherwise — we face the double bind exercised on much too low a level: the pupils are not offered responsibility at a level where there will be obvious demands on them beyond the reach of the "consultants".

In that case one risks the young people's chance of taking charge of their lives in a way which is inherent to the Activity concept being reduced and very little being achieved. The level of responsibility offered to children and young people is, in the end, politically determined. In the process of reproduction of society certain responsibilities will be offered while others are rejected. It is this which leads Henry (1963) to remark that if school really wanted creative pupils, it would throw doubt on the justification of the Ten Commandments and the two-party system. We do not usually question such institutions in school.

4.3.8. *Summary*

In this chapter I have discussed three theories in the perspective of Activity theory. They are dialectical theories which consider human behaviour as social behaviour. The dialectical nature of these theories demonstrates how human behaviour has to be interpreted *both* in terms of society and of the individual himself. Common to the theories is that they provide the educator with powerful thinking tools in cases where the learning behaviour of the pupil is obviously not as expected.

Mead's conceptions of *rationale for behaviour* and *Generalized Other*, Freud's discovery of *repressed knowledge*, and Bateson's conception of *metalearning* and *double-bind* are all useful for a full understanding of learning behaviour.

One critique against the theories and my interpretation of them for the purpose of education so far may be that they connect the individual *too* tightly with his environment, leaving him with few choices and possibilities for initiative on his own behalf. This is made clear by Bateson who connects the individual indissolubly with his environment: the human mind is immanent, but not only in the body. It is also immanent in pathways and messages outside the body; and there is a larger Mind of which the individual mind is only a sub-system. Any dialectics has to be examined carefully in each case it is observed. If A relates dialectically to B, both A, B and A ↔ B have to be analysed. Furthermore, if the state of affairs is not as desired, an analysis of where support is to be provided is warranted: A, B or A ↔ B.

This may imply situations where the focus should be on the pupil: "Look here — pull yourself together".

Neither Activity theory nor any other psychological theory can disregard the human ability to take initiatives, create and invent possibilities. A pedagogy, included the one I have developed so far, cannot be based on the design of an appropriate learning context with a careful eye on the frame-factors and leave it at that. The individual as an active individual must also be expected to take initiative in relation to this situation and utilise it. This point is not treated in any detail in any of the three theories analysed in this chapter.

What has been offered are thinking-tools which help to understand the pupils' predispositions for learning. The above indication that one should not disregard the expectation that the pupil will take the initiative will to some extent depend on the *level* of the double-bind.

The motivation for studying the theories of Mead, Freud and Bateson was rooted in the experience of a contradiction between the learning potential of large groups of school pupils, and their actual learning of mathematics as the subject was taught to them. Another reason for the feasibility of the theories is the pupils' ever present evaluation of the curriculum, in several cases resulting in ignorance, rejection or confusion.

My paradigm for building a mathematical education including these pupils will be to start with their resistance, taking advantage of it, rather than disregarding it or moralising. It is in this way that I *politicise* mathematics education.

CHAPTER 5

POLITICISING MATHEMATICS EDUCATION

> A society that frustrates or alienates a sizeable proportion of its inhabitants can survive only as long as it is possible to keep the discontented docile and to isolate or incarcerate those who refuse to be "rehabilitated". The helping professions are the most effective contemporary agents of social conformity and isolation. In playing this political role they undergird the entire political structure, yet they are largely spared from self-criticism, from political criticism, and even from political observation, through a special symbolic language.
>
> Murray Edelman

5.1. ON IDEOLOGY, HEGEMONY AND RESISTANCE

5.1.1. *Mathematics Education Is Political*

I shall here look at the failure in learning mathematics as a result of the pupil's lack of appreciation of the thinking-tools of the curriculum. In some cases this failure will be due to conscious resistance as the result of open rejection of the subject. In other cases the failure may be due to various forms of double-bind as the pupil does not have the appropriate metaknowledge of the conflict inherent in the messages sent to her through school.

In every case I shall consider such failure as political: some pupils are prevented from an important field of knowledge because of the design of the curriculum or the mechanisms of the examination system. I shall thus argue, with reference to my conception of what it is to be *political* developed in §1.2.3., that mathematics education, like other education, *is* already political.

The implication of this view must be that one should take advantage of the existing politics of mathematics education by utilising it for the purpose of curriculum construction. This means being open about the politics of the education, rather than disregarding this aspect when planning the curriculum.

Gerdes (1985) reflects the political aspects of mathematics education (for which I am asking in the case of curriculum development) in Mozambique.

By the *politicisation* of mathematics education I thus refer to the inclusion of reflection on the political dimension.

I shall thus have to explore the political nature of mathematics education, and thereby see how we can reach those pupils who have ceased to learn our subject or who never got the idea of learning it at all. As Willis (1981) stresses, if reading scores (or mathematics) are to exist, the authorities should constantly be attacked and asked why such scores are lower for working class and inner city schools.

I am thus disregarding pupils who may have specific handicaps for the learning of mathematics for whom a diagnostic label may be appropriate. It is, however, relevant to examine to what extent pupils with obvious learning handicaps will require thinking-tools which can be of specific use to them. I will not enter into such an analysis here.

5.1.2. *Reproduction of Society*

I have already referred to theories of social reproduction (§3.3.). Such theories describe the processes by which a society, explicitly or indirectly, develops and survives.

The underlying thesis is that any society has developed a particular mode of production, by which it has learned to exist in its world, and this mode of production develops various attitudes, strategies, ideologies etc. among its members. If such reproduction is based on a strong element of worship of a certain individual (a general, president or bishop), social reproduction theory will explain this and show how this worship is transmitted to the next generation.

If the life of the society is based on the oppression of one group by another, reproduction theory will also describe how such oppression is exercised.

Louis Althusser is a French structuralist whose analysis of the the repressive forces of the State and the resulting production of ideology is carried out in the Marxist tradition. Althusser (1972) distinguishes between the Repressive State Apparatus (RSA) and the Ideological State Apparatus (ISA). As the RSA rules by force by means of institutions like the army, police and courts, the ISA rules through consent: family, schools, mass media and so forth. Althusser gives school a predominant place in the ISA in the industrialised countries.

Another French sociologist, Bourdieu, has built another theory of social reproduction (Bourdieu 1977, Bourdieu and Passeron 1977). His works also have a clear affinity to Marx, although he does not follow Marx in the question about class struggle as the major drive for history.

Bourdieu describes how cultural reproduction represents the transmission of the culture of the dominant class, oppressing other cultures, creating the *cultural capital* of the dominant class. This implies that the cultural knowledge of the dominant class becomes the culture per se within the various state institutions, such as school.

Such use of cultural capital at the cost of those who do not possess it Bourdieu calls *symbolic violence*. This conception links up with the violence which occurs as a result of double-bind situations: the individual knows that he has cultural knowledge, but is rejected by those in a position to claim that there exists only *one* culture.

Today we can see how Althusser, Bourdieu and the rest of the social "reproductionists" of the 1965—1975 period failed to develop theories for Action. Their theories provided poor guidance for the educationists. Bourdieu comes out best with his conception of *explicit pedagogy*, which means that one should expose and neutralise the use of cultural capital in schools. This leads back to the double-bind structure: the necessity to communicate on the metalevel in order to reveal any contradictory messages in communication to the pupils. Still Bourdieu lacks a theory of Activity or of politicisation of education in general in order to reduce symbolic violence.

Less good is Althusser who, according to Giroux (1981a), reduces the "constituted subjects" to those helplessly trapped in the prison of the state apparatus, and as a result

> the conditions or even possibility of transcendence get lost in a grimly mechanistic notion of social reproduction.
>
> Ibid. pp. 5—6

Willis (1981) joins in declaring that

> In an awesome reverse of the Medusan myth, Reproduction theorists look back to cultural production and turn it, not themselves, into Stone.
>
> Ibid. p. 49

I do not want to be so harsh. In my Introduction I have tried to draw a picture of some of the confusion, curiosity and conflicts of the 1965—1975 period experienced by intellectuals. It was a period when several educators were searching for an understanding of the foundations of society. In that respect Althusser, Bernstein, Bourdieu and the rest of the 1965—75 period held an important position. After analysis and reflection, Activity has to come. Thus we have to look in new directions.

5.1.3. *The Pupil as a Purveyor of Ideology*

In §4.1. I analysed the concept of *ideology* in the context of the Generalied Other. I conceptualised ideology as the set of attitudes which the individual takes over from the system of groups she has as referents for her behaviour. I also argued for the extension of the GO-structure by including both its historical and material frames. I pointed to the possibility of conflicting GOs, which will imply a corresponding "conflicting" ideology, that is, an ideology may contain attitudes which conflict with each other.

Clearly such a conception contradicts most of the previous conceptions of ideology. Usually ideology has been studied in the perspective of the *production of ideology* by various social classes (usually the dominant) and what Althusser calls the ISA. In education we have mainly seen studies on the ideology communicated through the curriculum by means of textbooks, teacher education, examination systems and so forth.

I shall now analyse the GO-concept of *ideology* a little further. First of all, what has been said about social reproduction theory indicates that there can exist hegemony of one ideology over another as one can dominate another by the use of power. In the case of mathematics education this point is easily demonstrated. We need only refer to the person who is in a position to define mathematical ability or the mathematics curriculum, and examine what kind of attitudes such definitions create towards various groups of pupils.

My text so far is full of examples of pupils who are not usually recognised as able mathematicians by those in charge of a school. Yet I will argue that their potential for mathematics in one form or another is unquestionable. Conceptions such as *Folk Mathematics, colloquial mathematics* and *Ethnomathematics* fully demonstrate the point: mathematical knowledge exists in many forms.

I do not argue here for conceptions of mathematics different from those most prevalent. My point is that even a conception of a "neutral" subject such as mathematics will not escape an ideological stamp if it is analysed in detail.

The concept of a *dominant* ideology is usually connected to *universalism*. Keane (1983), Giddens (1983) and Thompson (1983) all take the position that ideology has to be defined in terms of hegemony, as political belief systems which by their structure appear to be universal:

> The concept of ideology might be applicable to any and all sectional forms of life which endeavoured to represent and secure themselves as a general or universal interest; ideological forms of life are those which demand their general adoption, and, therefore, the exclusion and/or repression of every other *particular* form of life.
>
> Keane, op. cit., p. 16

We see the connection with psychoanalysis: it is exactly those attitudes which by dominance can appear as universal which cause the conflicts resulting in repression (Freud) or double-binds (Bateson). Gramsci (1971) refers to *contradictory consciousness* in such cases. He thinks of how individuals view their world from a perspective that contains both hegemonic forms of thinking and modes of critical insight: the worker gains wisdom both from the newspapers and the shopfloor.

The educationist now faces several problems. First of all we have to be concerned about the fact that every pupil carries some ideological structure, however contradictory it might be, a structure which is not necessarily one which claims universalism. The phenomenon of pupils' ideologies relates to the phenomenon that school represents ideologies as well. In the terminology of the GO: school as a GO represents certain attitudes about right and wrong. This connection, pupil-school ideology, causes Giroux (1981) to argue that a concept of ideology has to be developed that provides an analysis of how schools sustain and produce ideologies as well as how individuals and groups in concrete relationships negotiate, resist or accept them.

By this he also indicates the next point which the educationist has to consider: the pupils are there, even those who do not learn. In Giroux's words: they negotiate, resist or accept. I would add, some also withdraw, become passive and turn silent. Still, the pupils exist.

It is here Giddens' concept of *structuration* becomes important.

Whereas Althusser describes how we live *in* the ideologies, Giddens claims that we also *create* the ideologies as we live in them:

> Actors are always knowledgeable about the structural framework within which their conduct is carried on, because they draw upon that framework in producing their action at the same time as they reconstitute it through that action.
>
> Giddens, 1979, p. 145

In many houses the housewife still does not escape her kitchen, whatever she may think about it. The worker has to stay at the assembly line on the shopfloor and has to adjust to it. The pupil still has to start her schoolday with a psalm, sit in a classroom which is furnished in a particular way and so forth. They all develop attitudes as a result which may or may not contradict the attitudes derived from other social settings. There exist both a production and a reproduction of ideology:

> Ideology is more than the reifications of socioconsciousness and social relations, it is also consciousness struggling to constitute itself against the objectified nature of social life.
>
> In other words, human beings in the course of their work and everyday lives are never reduced to the objective representations of reified social order.
>
> Ibid., p. 20

Ideology is both a mode of consciousness and practice, related to specific formations and movements (Giroux 1981). Ideology thus creates and shapes Activities, and is also transformed by these.

Structures can develop *resistance*. If a dominant ideology *x* rules over ideology *y*, thus having the power to appear as universal ideology, it is hard to see how ideology *x* can invite the *x*-individuals to Activities supporting *y*. Such an attempt would imply double-binds or contradictory consciousness.

A vital task for a non-oppressive education will be to focus precisely on the field of tension between such an *x*- and such a *y*-ideology. The motivation for the exploitation of psychoanalysis was its stress on the phenomenon that oppressed experience does not disappear. It rather repeatedly turns up as forms of *resistance*. Such resistance proves the inadequacy of the social reproduction theory of the 1965—1975 period. Willis (1981) shows how it is impossible to fully understand social reproduction without understanding something of the nature of the groups whose material presence constitutes the reproduction.

In other words: social groups resisting social reproduction, such as those rejecting school, are real. Their existence has to be included in theories about such reproduction. It is such resistance which proves indispensable in treating the politicisation of mathematics education methodically.[1]

5.1.4. *Resistance*

To locate *resistance* I need a concept of *culture* in which the resistance is produced. A political concept of culture must include the study of how people live their lives within a particular structuration. This includes the ideology concept as well. Such a concept of culture is thus not a purely anthropological concept which applies descriptively to the arts, leisure and work. It is rather a concept which claims to examine the relationships between the social construction of the individual and the way she lives her life.

Culture is thus not simply the lived experiences within society, but the lived antagonistic relations

> situated within a complex of socio-political institutions and social forms that limit as well as enable action.
>
> Giroux, 1981a, p. 18

Examining *music* as a political cultural phenomenon will not only implement the description of various forms of music within a society and the distribution of these forms, but also implies the analysis of how music relates as a lived reality to that society. In particular, it will be important to see the *function* various forms of music have for different groups of society as related to their Activities.

The political component of such a concept of culture has been applied in drug education. An informing study of a *drug culture* has to emphasize the function of drugs. The educator/parent/therapist can hardly ask a consumer of drugs to cut her use without knowing the relationship between the drug and the user in terms of *function*.

Similarly *mathematics* is used in cultures and is influenced by culture. The existence of Folk mathematics demonstrates that mathematics is both produced and reproduced: mathematics is used in the daily life of people, and thus mathematical knowledge is produced — and the way mathematics is used depends on how mathematics was previously developed. The same dialectics naturally exist in mathematics research. As in the case of drug education it seems profitable to ask about the *functions* of mathematics in a particular culture (as in some Folk mathematics or at the university institute), in order to consider the political (and thus cultural) aspects. When designing a mathematics curriculum in an area where the culture is related to, say, timber production, the argument will be that it will not only pay to relate the mathematics to timber-mathematics, but also to relate the educational tasks to the function of mathematics in timber production.

Such a political concept of culture is dialectical as it sees the individual as both a producer *and* reproducer of culture. What we experience and will take advantage of in the context of education is the existence of resisting cultures and counter-cultures which reject the dominant ones.

The phenomenon of resistance as it can be observed in various forms and at various levels sustains human dignity and rejects the oppression which dissolves personality and buries the potential for Activity. It is mainly the production of such cultures that Willis's (1981) works are about. He describes cultural production as the field of creative self-making in a subordinate class. My aim is to catch resistant cultures in their developing phases, accepting the signs of rejection, eventually turning them into Activities which can challenge dominance. The argument is that the thinking-tools of mathematics can be indispensable to such a challenge.

I deliberately use the notion of *challenge*. It includes the possibility of reducing dominance, eliminating it, the vital point being that a challenge makes a conflict or a contradiction visible, and thus easier to relate to.

I hinted at a project in §1.2.7. I shall now describe it in detail. I am in a class in a suburb of Bergen. The kids are aged 11—12. Their teacher describes the typical pupil of this class as living alone with the mother and another child. The father will sometimes bang on the door to get in. We understand this is not a school in the most prosperous area of Norway.

In one lesson the pupils were to do a project: do you know yourself? Write three positive things about yourself, and three negative things.

They sat in groups, and Kjetil was the first one to say:

"This becomes difficult. It will be easy to tell about the negative ones. But the positive things?"

Agreement in class.

I, an intruder in their class, said to them sometime later:

"I don't like it that a bunch of great kids such as you cannot say anything good about themselves."

I know something about their culture, so I introduce the notion of *responsibility* into the discussion. They have quite a lot of responsibilities these kids. Being alone in the evening, going to bed without looking at the video-nasties which are very popular. Buy the food, take their sister to the kindergarten, take care of mother when she is dead drunk. The latter was translated as "take care of adults when they are sick."

We had some material. I wrote the word *documentation* on the blackboard, we discussed its meaning, and I demonstrated some thinking tools from statistics to them: the mean, the mode, the median, percentages and so forth.

"Now let's document the responsibilities of this class. Next time someone attacks us, we can document: Don't say that we don't do anything for other people."

Documentation was produced, and presented in oral and written form in class.

Next project: Decisions. When do we participate in family decisions? Many teachers objected to this project. It is intruding in family matters. I do not know so much about the trend in other countries, but in Norway schools' concern with the social aspects of upbringing is continually increasing. So in this particular class, belonging to this particular culture, the project could cause no trouble.

So, documentation. How?

They had collected material about the fields and frequencies of their participation in decisions. And they were to document it.

"Tell us how, Stieg." Stieg wouldn't. Astonishment in class.

A whisper goes around. "He does not want to tell us."

"No. You have the thinking-tools. You have to decide yourself how you will use them."

Class becomes a bit noisy. Their teacher is very restless. "They will never manage."

20 minutes, long as years. Then the first group started. The other groups followed.

Some time later they finished. And — I refer to Chapter 2 — as expected, the groups had documented different variables, and they expressed their results in different words. The next step in such a chain is that the pupils face some situation where documentation is needed. And they apply the tools themselves. This situation is hardly the task for school. It is rather the aspiration and hope of the educator.

* * *

I was 72 degrees North, visiting a small town. There had been some damage at the largest secondary school (an immense monstrosity of a building, a result of the period of school centralisation). A teacher had been beaten up by some of his pupils. I had been invited by the Teachers' Union to talk about Discipline. I had to use three aeroplanes to get there, each smaller than the other, and reached ice and a completely dark day. In the morning I visited the comprehensive. My host was an amiable, radical chap, who worked in a tremendously stressful situation. Oslo was far away, and so was the pupils' interest for the State curriculum of Norway.

Entering the classroom I found them at once. They had black leather jackets, studs, silver bracelets and necklets, and tiny pointed shoes. I moved up to them. They were to do written exercises in grammar, converting sentences of direct speech into indirect.

It did not turn them on at all. A pencil was sent through the air, a shout about yesterday's events, and perhaps a sentence or two were written in the book.

"It is all a lot of rubbish", they said.

I gave them some support, and they hesitated and became interested.

"These fucking books. We'll burn them. We'll make a big fire of them in the yard in June."

I still did not oppose their point of view. Actually, I agreed.

"Look at that. It reads 'Have you ever been to Lillehammer?' "

He looked at his mate. Lillehammer was 3000 kilometers away. He shook his head. He never had, but it was in his book. And he had to transform the sentence, that was what was written in the book.

They looked at me. I agreed again.

"It is rubbish, most of it."

It was so easy. My poor colleague was running around, helping those who had started to work, while the rest of the class made the most of it.

I had no responsibility for the class and suggested:

"You should collect some statistics. See what it is in this book that makes sense to you. You could collect some data about it, and report back to the publisher. Ask them why they publish such rubbish".

"They wouldn't answer."

"Yes, they would. If not, you press them. And you report to the local authorities, asking for different books next year."

"Oh, they wouldn't listen to us."

"Then you make them listen."

I suggested it to my teacher friend. There were both mathematics and the mother tongue in it. He looked at me as if I had asked him to revise the complete educational system on his own. He could not do it. Perhaps he was right.

In other fields, however, he and his colleagues had initiated projects which developed Activities in every sense of the word. The sad thing is that mathematics is not usually regarded as useful for integration into such projects.

In what follows my message will be about how to convert passiveness, indifference and destructiveness into constructiveness through Activities. It will be a message about *accepting* resistance rather than rejecting it. Mainly the message will be to treat the various thinking tools of the major subjects, especially those of mathematics, for the purpose of such Activities.

Frequently educators can learn important instructional strategies from professionals working outside school. The leader of the Street Section of the Social authorities in Bergen, Gro Lie (1981), talks about the x-gang which caused embarrassment to old people and car owners in their street. The gang never used violence but there were incidents of abusive language, and they just hung around, *being* there. They would get pretty close to the cars as well, perhaps leaning against one or two.

The social workers of the Street Section came into contact with the gang, analysed the problem, and found it to be a familiar one: they did not have anywhere to go in the evenings. What followed on the initiative of the Street Section is a demonstration for any educationist of how to initiate some Activity, thus turning destructiveness into constructiveness. Having analysed the situation together with the gang, they searched for a suitable place to use. They then lobbied for their plans, put some pressure on the authorities when they turned their backs on the plans, and also tried hard to establish a good relationship with the adult population of the street. The tools developed were mainly in the field of literacy: preparing for a press conference (role play), preparing and arranging a public meeting (with the inhabitants of the street), preparing for meetings with the authorities, and so forth.

Characteristically, hardly any thinking-tools from mathematics were applied. This failure to use mathematics in important Activities is usually due to either (a) the educationists' lack of knowledge of the potential of mathematics or (b) the mathematicians' lack of knowledge about the Activities of ordinary people for which mathematics is needed.

5.1.5. *Activity as a Drive for Ideology Production*

Going back to 72 degrees North, we will find that the kids up there are just as interested in football as they are elsewhere. There are no grass fields, and it will be dark in the daytime in late October. But the school is huge and built around a large hall, the size of a handballfield. So there is a winter tournament for 7-a-side, each class fielding their team consisting of 10 players. Lunchtime is matchtime. The problem is that the teachers state as a condition for the tournament that there shall be equally many girls as boys playing. Can we hear the boys murmuring? Can we hear some remarks about "The silly girls will ruin the game completely. We are better off without them?"

See how the temptation to play in the tournament is the stronger feeling, and watch the girls as they enter the field with the boys! Can we see how they get no passes from the boys at the beginning of the match, until the boys discover that this is not the best tactics in the world if they want to win? Can we see that there is a slight change in attitude as the tournament continues and the girls prove themselves? And what about the use of thinking-tools?

After my observation of these activities I would try some of the following tasks in mathematics:

There are laborious tables to work out, not to mention the fixture list. It was an "everyone against everyone" tournament, and the match days had to fit in with the timetable of each class. The resulting combinatorial problem is quite complex. Could it be solved by means of a computer?

And what about statistics in order to focus on the sore point: the girls' contribution to the game as the tournament went on? What about the number of decent passes in the end as compared with the number in the beginning? What about the number of passes they received from the boys and the resulting score? Which thinking-tools from mathematics will be suitable?

My point is that ideology will be transformed mainly as the result of Activity (which I claim this tournament was for several of the boys and girls).

We notice the presence of *inertia* among the boys who have to give part of their field up. It is resistance the other way, to protect hegemony.

The girls are not to be stopped: in Norway they have their own national league and they referee men's matches. A female reporter reports football on radio. The female sports reporter of the Norwegian National Television counts the days until she is to report on the cup final.

It is by contesting hegemony, not only by talk, but also by Activity, that ideologies can be challenged and transformed.

5.1.6. *From Critical Awareness Towards Activity*

There still remain several problems about *ideology*. What about the situation that usually faces the teacher: conflicting ideologies are present within the same classroom? What about reactionary ideologies, such as those favouring sexism and racism? Willis (1981) points out that such ideologies had a strong influence on the working class boys among whom he did his research. The phenomenon is not restricted to his sample: the present sympathetic views among young people about the various facets of National Front ideology witness the same.

Willis faces difficulties here when he makes an attempt to devise educational strategies to overcome such problems. So does Giroux. Although I can hardly claim to have solved the problem of transforming reactionary ideologies, I cannot see how we can avoid a consistent and forceful application of the Activity concept in order to approach a solution to this problem. The paradigm which directs my thinking is that Activity is a decisive force for the *transformation* of ideology. Ideology guides Activity, but the development of ideology is in the main a result of the Activities of the individual and his group. Examples of how Activity may challenge prevalent ideology will be reported in the next section.

Giroux (1981, 1983) has, with his books, brought the theory of the politicisation of education a big step forward. But while Giroux takes us to the door facing an educational praxis, he does not take us through it. He does not ask his reader to do so either. Giroux apparently does not regard School as a camp for Activity.

He restricts his theory in such a way that the educational implications of it are reduced to a critical awareness of the oppressive forces in society. To the extent that is true, he reduces an educational praxis to the praxis of the psychoanalysts as described in §4.2.: the subconscious will be disclosed, but nothing more than that. It seems that Giroux restricts education to be about thinking, in particular critical thinking leading to political awareness, thus excluding Activity:

> The foundation for a progressive form of classroom pedagogy can now be developed since schooling can be understood as a study in ideology and values.
>
> Questions concerning totality, mediation, and appropriation now become essential

> in developing a form of pedagogy in which teachers carefully examine how the structural and ideological determinants of the dominant society affect the behavior, attitudes and speech of all those involved in the classroom encounter.
>
> Giroux, 1981, p. 123

The same restriction turns up in Giroux's critique of Freire. Giroux quite rightly (according to Activity theory) criticises Freire for treating communication without regarding its ideological superstructure:

> ... communication which is freed from domination becomes only meaningful when measured against socio-political arrangements next to which all institutions and social relationships in the larger society must be measured. By doing this, Freire ends up with a pedagogical model for communication that appears abstracted from institutionalized discourse and social relation in a larger society.
>
> Ibid., p. 138

Critical awareness of society has, according to this, to include the awareness of the political uses of language. In the case of mathematics education this would imply the awareness of the ideological content of mathematical models: what is being left out in the mathematical model, and which economic, physical or social theory leads to the relationships which are mathematised?

In sum: What is the ideological foundation in the textbook which designs a project about income and wages in a secondary school?

And again: What is the ideological foundation for the mathematical models exploited in studies of strategy in the Army and the Ministry of Defence?

Mogens Niss (1983) has built a complete curriculum for teacher education based on such critical awareness of the ideological foundations of mathematics. His students examine mathematical models used by the government. They persistently ask why and how x and y were related to each other, and why $x, y, z, \ldots$ were not related in other ways. Clearly there are political, i.e. ideological reasons, why x and y were related, and other relationships were not introduced into the model. Following the encouragement of critical awareness, or better, together with critical awareness, Activity, that is, political action should come.

Both Giroux and Freire ideologically regard Activity as a necessary and desirable step following a critical awareness of society. In theory they do not see this as a task for the school to initiate. One problem of such a position for the teacher in the classroom is a problem of *faith*: "See Teach, all this talk, it is certainly right, but what then?" It is hard to see how pupils who may be in a difficult life situation can have faith in their future curriculum if their metaconcept of it tells them that school helps them to understand what is wrong, but does not stand up for them in a common attempt to challenge the difficulties. A call for a critical awareness excluding action may be interpreted as just another discussion, which will have been heard too many times before. There is a profound metalearning in young people who are fed up with those who can talk about society but do no more than that.

It is difficult to ignore the fact that Activity is necessary for the development of a critical awareness. Activity is about the transformation of the physical, social and political environment. Activity thus leads to new experiences, confirming and questioning former attitudes, awareness and ideology. Willis (1981) suggests the need for Activities which resist cultures in the context of education.

> Even the anti-mentalism of resistance groups might be overcome where cuts threaten, practical classes, games, and clubs in which their cultures thrive more happily than in stricter academic classes.
>
> The question is one of linking general principles to a flexible practice.
>
> Ibid., p. 66

Willis furthermore emphasizes the educator's responsibility: the problem is now not to ditch theory but how to reach for this theoretical possibility in practice. If theory should now take note of practice, then practice should take note of theory. To restrict education to the promotion of critical awareness is in my opinion to make it too easy for the educator. Nothing is easier than to discuss with the pupils how bad (good) their state of affairs is and to speculate about the reasons for this. It is much more difficult to follow up and test the content of the discussions by praxis.

Do we remember David Hansen? The teacher with the traffic project? (§1.3.) Obviously he had a project on his hands which could easily lead to political action. The documentation which the pupils of his class could collect about the heavy traffic driving through their playing area, could be followed by addressing the authorities. If nothing was done by the authorities, still important political knowledge would be gained.

In common with Giroux, Willis and most sociologists of education have completely neglected the role of thinking-tools for critical consciousness. They all join the large group of pedagogues who claim that pedagogy is about human beings, not subjects. Willis may refer to the need for the initiative of the practitioners here. Still it is surprising to see how some of the deepest thinkers in the field of sociology of education to some extent disregard the functionality of knowledge as a means for a democratic education.

It is a reasonable counterargument to my critique here that as school is the most important tool for the ISA (as described by Althusser 1972), one can hardly expect the same State Apparatus to tolerate not only a critical awareness of possible oppression exercised by it, but also tolerate Activity opposing it. Still, in several places, practice has proved that such Activities are possible, within certain limits and within certain fields. A way of explaining this is to relate such tolerance to contradictions within the State Apparatus itself. As economic depression develops and life conditions deteriorate accordingly for a growing group of people, the firm grip of the ISA may be weakened. A reason for this is the weakening of the ISA as inner conflicts grow as the depression grows. The major limits for Activities in the educational context seem to be linked to the capacity of the educators. This weakness relates to several circumstances. One of these is the lack of insight into the life situation of large groups of pupils which creates a need for the methods of social anthropology in education.

A difficulty which is related to mathematics education in particular is the lack of experience of how to apply the thinking-tools in such a way that they are recognised as functional knowledge by various groups of pupils. We have, however, some experience to build on. Some educators have begun. We shall learn from them, improve what we learn and make our praxis grow.

5.2. FROM CRITICAL AWARENESS TO ACTIVITY

Individuals in many schools, colleges and universities are beginning to develop learning materials and styles of working which encourage more active participation in decision-making on the part of the learner. Making choices and commitments in real world context should develop from, and be linked with, decision making in the classroom.

Editorial *Contemporary Issues in Geography of Education*, Vol. I (1) 1983.

Only when we are clear about the kind of society we are trying to build can we design our educational service to serve our goals.

. . .

It has to foster the social goals for living together and working together, for the common good. It has to prepare our young people to play a dynamic and constructive part in the development of a society in which all members share fairly in the good or bad future of the group.

J. Nyerere, *Education in Tanzania*

5.2.1. *Conscientisation*

I have spread parts of Freire's works throughout this book, and will here restrict myself to a short summary of the major conceptions. (See also §§1.2.2.—1.2.5., §3.1.5., §4.2. including Note 1 and §5.1.6.)

A. Integration and adaptation

Integration with one's social environment as distinguished from adaptation is a specific human ability. Integration emerges from the individual's capacity to adapt to reality and the capacity to make critical choices and to transform that reality. To the extent that man loses the possibility of making choices and transforming reality he is no longer integrated, but adapted to society.

Integration can here he read as "to be permitted Activity".

B. Conscientisation (conscientizacao)

Conscientisation represents the awakening of the critical awareness of the individual. It will not appear as the result of economic changes, but must grow out of a critical educational effort based on favourable conditions. It is a process in which man discovers himself in his cultural setting. Culture is here to be interpreted as political culture (§5.1.4.).

It is a criticism of Freire's views, as seen in the perspective of Activity theory, that conscientisation, although it is directed towards Activity, is conceptualised independently of it.

C. Literacy

Literacy is an indispensable field of praxis for conscientisation and integration. In the culture of silence, the masses are "mute". They are exempted from creatively taking part in the transformation of their society, and therefore exempted from integration — Activity. Freire emphasises that the literacy process must relate *speaking the word* to the transformation of reality, and to man's role in this transformation. It seems, based on his praxis, that he by this restricts himself to the principle that knowledge becomes liberating when it "can be released from reifying social and political relationships" (Giroux (1981), p. 131).

Activity theory would say that this is a necessary but not sufficient step towards Freire's goal of integration. The next step would be to initiate Activity.

D. Knowledge is social knowledge

For Freire knowledge is social knowledge. Knowledge does not exist in people in their world, but *with* people *with* their world. Knowledge exists by means of dialogue as shared knowledge. This links to Activity theory: Activity theory consists mainly of collective actions; individual actions exist in the context of collective efforts:

> — the role of man was not only to be in the world, but to engage in relations with the world — that through acts of creation and recreation, man makes cultural reality and thereby adds to the natural world, which he did not make. We were certain that man's relation to reality, expressed as a Subject to an object, results in knowledge, which man could express through language.
>
> Freire, 1973, p. 43

It is one of my main goals in this book to argue for a similar rôle for mathematics to the one Freire designs for the spoken and written language.

5.2.2. *The Cultural Circles*

The problem of Freire's theory and practice is demonstrated by his major form for praxis: the cultural circles (ibid.). These circles were developed according to theory: school in the traditional sense was regarded as a passive concept. In the circles there would be no teacher, only a coordinator. There would be groups of participants rather than pupils. Finally there was no alienating curriculum, there were "compact programs that were broken down" and "codified" into "learning units" (ibid., p. 42).

Freire points out that knowledge without action is meaningless: in the cultural circles attempts were made to clarify situations or to obtain action through such clarification. Here we encounter difficulties in following Freire. Does the action belong to the circles, or is it an appendage to them? Freire says that the topics for discussions were offered by the groups themselves. As examples of themes which repeatedly appeared were nationalism, the political evolution in Brazil, the vote for illiterates, democracy, etc.

Political action as a result of political awareness followed by literacy? Maybe. It is, however, difficult to see how such nationwide political problems can lead to political actions, i.e. Activity, as initiated by conscientisation developed as a result of the cultural circles.

I would argue that the topics of Activities would, in their initial stages at least, be found in the nearest social environment of the members of the circles. It is hard to see how the power of the ability to speak the word in connection with political awareness can be demonstrated on the level of "the political evolution of Brazil" so that it should imply political action. My doubts are mainly due to the importance of the mute individual's needs to experience herself as one who has a potential for Activity, an experience which will probably be easier in lower levels of ambition than those referred to here. I would support political action in the nearest social environment of the individual, leading, if this is wanted, to grander projects.

The confusions increase when examining the drawings Freire (ibid.) presents as "situations" discussed in the circles. There are ten altogether, picturing respectively:

- Man in the world and with the world, nature and culture;
- Dialogue mediated by nature;
- Unlettered hunter;
- Lettered hunter;
- The hunter and the cat;
- Man transforms the material of nature by this work;
- A vase, the product of Man's work upon the material of Nature;
- Poetry;
- Patterns of behaviour;
- A cultural circle in action — synthesis of the previous discussions.

It seems to be a long way from these topics to the political problems of Brazil. On the other hand, the above situations invite an awareness of the social environment, including its historical and political components, such as I called for above. I think that such situations as Freire constructs here are vital for a critical awareness of the specific culture, which in its turn is vital prerequisite for Activity. They relate to the kind of Folk Mathematics I discussed in §§1.1.2. and 3.3.5. for some of the same purposes: the pupils' critical awareness, in particular that of the female pupils, that they possessed important mathematical knowledge not so far recognised by school.

Freire's further programme consists of several phases. The first two comprise research into the vocabulary of the groups he is working with. The next is the selection of generative words from this vocabulary, which in the end leads to the fifth phase: the preparation of a set of cards on which the phenomic families which correspond to the generative words are printed (ibid., p. 52).

Conscientisation by means of mathematics: The moving project
Norway, Class 1 (7 years old).
To understand this example one must take into consideration that Norway is a country with a scattered population. A move from one part of the country to another is still not common among the population. Recent changes in production have altered this.

Fig. 5.2.1.

"Miss, we are going to move again".
"Oh, I am sorry Sue. When?"
"Next week."
"Miss, we will probably move next month."
Miss nods. It is the morning session, she has the plan for the day ready, but takes time to show interest.
A week later she is prepared.
"How many places have we lived in?"
Each pupil contributes a number.
In a short time the class has a sample of 26 numbers ranging from 1 to 8. The mode is 4. Altogether the pupils have lived in 98 places. The 8 who have lived in 5 or more places total 62. Lots of calculations are done. Each result is derived from the figures which the pupils have produced, figures which are very important to them in the sense that a move from one place to another is an important event in their lives.
Frustrations, excitement, sorrow, fear of becoming lonely are some of the words for it. But:
"Why do we have to move several times?"
The pupils give different answers.
"Because father got a better job."
"Because Dad lost his job."
"Because Mum and Dad wanted to go to college."

What is the most common reason for moving? The children do not agree upon this. They have different ideas about this according to different experiences.

The teacher contacts a colleague who works in the countryside. Could she do the same project? Interchange of material. Comparisons. Can we communicate our findings to the other class? What were our figures? What are our ideas about this?

The children talk of their own experiences. And their arithmetic is founded on this.

The above story is not exactly as reported here. It was somewhat more complicated. There were five experienced teachers attending an in-service course in mathematics. They were very cautious, as they believed they would have problems in getting something out of the mathematical didactics. They worked in five different schools in the town and the countryside. And they tried this project in their five classes, communicating as outlined above.

5.2.3. *Between Conscientisation and Activity*

One of the few contributions offered on the question on cultural mathematics as a basis for practice is given by Fasheh (1982). He draws from his experiences as a mathematics teacher on the West Bank among Palestinian Arabs under Israeli occupation. He describes a first grade teacher making a chart together with her pupils for keeping a record of absentees. The approach is similar to the one I described in §2.2.

After a month the distribution can be studied. It turns out that the biggest number is on Saturday, and the question is why. Several possibilities are discussed in class.

- It comes after Friday. Friday is the official weekly holiday.
- The kids like to spend the Saturday at home because their fathers are there. Some of the fathers from the village work in Israel and so Saturday is the day off.
- There is a poor transport service on Saturdays due to the fact that some workers don't go to work.

Not much imagination is needed to understand that such statistics will stimulate discussion about important factors in the children's lives. As Fasheh himself expresses it — the project deals with a problem that is familiar and interesting to the children simply because they live in it. Still, I am not completely satisfied, neither with Fasheh's project nor with several of our own projects reported so far. My hesitation relates to the critique of Freire's theory.

The projects represent a leap forward, as they draw directly on the children's experiences. The thinking-tools of mathematics are used for the purpose of gaining new information. So we may possibly obtain conscientisation. But what about Activity? Both the Moving project reported above and Fasheh's project exclude Action, as far as the descriptions here are concerned. At what level can we expect the 6—7 years old to participate and initiate Activity? When does conscientisation contribute to Activity? Is it possible to see young children participate in the politics of the family or in the politics on the West Bank? Or are the contents of these projects just making a difficult life-situation even more difficult for the children as they explore it?

My thesis will be that such projects are only valid in the context of Activity theory. Then they are positive for the children to the extent they contribute to their political life, integrating them into the politics of their social environment.

As I have hinted before, we lack experience of how to integrate the subjects into such educational work, thus leaving most of the real problems of life to the social pedagogues. The uses of mathematics together with the mother tongue in its written and oral forms should obviously be put in the hands of the children.

We do know that children have participated at high political levels in many places, such as in Afghanistan, Eritrea, and Soweto. We thus know that the potential for Activity including all the aspects of political action is present in children, usually much more powerfully than we realise as educationists. I am thus in search of both methodological and theoretical principles for implementing the school subjects in general, and specifically mathematics, into such political action. I think there is no other way if education is to contribute in an honest way to young peoples' growing political participation in society.

5.2.4. *Politicising Mathematics: Challenging Ideologies*

I am now with the observation class described in §4.3.3. This project was the first we did together. I had made some enquiries before I met them. I knew about the largest industry in the area — a bread factory.

I write to the union for the booklet showing the rates of pay for different sorts of work. This booklet, the *tariff*, tells us about the wages for skilled and unskilled workers, for different age groups and for levels of experience. It tells about overtime, 50%, 75% and 100%, related to specific periods of the day (night), Sunday work and so forth.

The figures are of the type 21.47 kr an hour, rather than 20 kr as in the text-book. It is clear that we will need hand-calculators, and it is equally clear that we have to use them critically by using mental approximations as a check.

My previous investigations told me that there were mainly four major discussions on the shop floor:

1. Why were those women who did the same job as the "skilled" men rated as unskilled workers?

2. Was it a good or a bad thing that personal benefits were offered to some of the workers extra to their pay? And why had such benefits to be kept secret?

3. Local negotiations were coming up. Should the workers follow the hard line or follow the employers' request for a modest claim ("You want your factory to survive, don't you?")?

4. The value of union membership.

Fig. 5.2.2.

I decided to focus on the first two discussions. The boys in the class were first presented with the leaflet from the union. The name of the factory was on its front page. It immediately awoke their interest, and we started to calculate wages for different categories of workers for:

— 1 week;
— 4 weeks;
— half a year;
— a year.

I produced a list of workers. I deliberately constructed the list for the purpose of initiating discussions:

Agnes Andersen, unskilled, 22.95 kr (per hour)
Merete Lien, unskilled, 21.93 kr

Nelly Foss, unskilled, 21.29 kr
Astrid Larsen, unskilled, 20.73 kr
Berit Holst, skilled, 24.51 kr
Linda Dale, 24.51 kr + 80 kr a month
Terje Olsen, unskilled, 22.55 kr
Arne Holen, unskilled, 22.55 kr
Jan Aasen, skilled, 32.45 kr + 120 kr a month
Kaare Nilsen, foreman, 4890 kr a month + 300 kr a month
Einar Dalen, foreman, 5240 kr a month + 300 kr a month.

Now the discussions start as hoped. The fulfilment of such a prophecy is of course completely dependent on the fieldwork. The boys had mothers and sisters. They had fathers and brothers. Some of these were redundant. Some feared unemployment. One mother had just come onto the dole as the industry she worked in (cleaning) had been laid off, not because it was not profitable, but because it was not sufficiently profitable for the owners. The boys had lots to talk about as they did their calculations.

Fig. 5.2.3.

The boys knew that although they had not been able to participate in events like those we had analysed, their turn would come and the discussions would be theirs. So just as the gifted pupils do not question their algebraic expressions because they know that some day they will probably be important to them in some way, these boys were grabbed by these matters for the same reason. The signs were the discussions. All the familiar arguments came up. I never stop being impressed at the diversity of the arguments and the various experiences the children build on.

Different views were discussed, different ideologies were expressed. While some kids would have kicked the women out of their jobs, the others protested: they knew by experience what it meant to have their mums on the dole.

Still, it was all just words. We could do nothing about the situation. We could explore it and inform each other, gaining some new knowledge. The ideologies were thus challenged. Perhaps one or two of the most ignorant boys revised their attitudes as they heard the experiences of the others. But we were not in a position to do anything there and then. We are preparing for what would come later.

This project provided the possibility of examining mathematics in the narrow sense. Distributivity is one example. As Agnes Andersen earns 22.55 an hour and Jan Aasen 32.45 an hour, it might be interesting to see what the difference would be in one week (40 hours), one month (4 weeks), in half a year and in one year. This can be calculated in two ways by means of the distributive law. That is, what about the additional 120 a month which was Aasen's personal benefit?

And what about the 100% for Sunday work, is this because of distributivity?

What about examining the function

$$\text{number of years in work} \rightarrow \text{yearly income}$$

for men as compared with women?

What does the average growth in an interval mean for this function, and how can we interpret the area under the graph?

Can we forecast anything about the future development of wages by means of these thinking tools?

Can we use the tools, i.e. the average growth and the area, as a basis for the discussion of women's wages as compared with men's?

Can the application of these tools sharpen the arguments, eventually be the basis for tariff negotiations?

Do we see other relevant possibilities for studying variations in the above figures? What about the pupils: It is possible that even these pupils will come up with suggestions about what to examine further?

In such a way mathematics obtains some legitimacy which leads to further steps into arithmetic and algebra.

Fig. 5.2.4.

I have already stressed that there is no reason to claim everything is rosy.

The boys had developed rough methods to survive as the result of hard and exhausting battles with the school. They contributed to our projects as described above. Only rarely did they work our way for the whole session. And they never worked overtime.

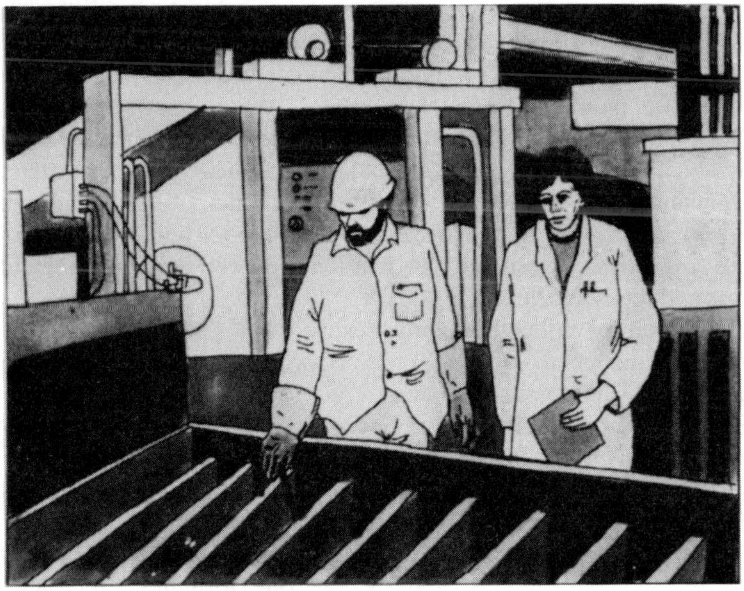

Fig. 5.2.5.

Investigations in women's participation in work outside the home usually go down well at all levels in school.

What is the proportion of women in skilled work, unskilled work, administrative jobs, in the unemployment statistics and so on?

What is actually happening — is it mainly women's jobs which are threatened today, and what are the arguments for that? What is the local experience?

5.2.5. *Health Careers*

A British sociologist, Nick Dorn, takes a step further towards initiating projects which call for Activity. He does not build any complete social theory, and neither literacy nor functional knowledge are his concern. He is a specialist on drug education.[1]

I refer to the analysis of *responsibility* (§4.3.6.) using the case of drug education: when facing drugs the pupils are on their own and thus have to be strong enough to cope with such situations. This includes the prophylactic aspects, the building of healthy social environments by and for young people.

Dorn's approach is to consider the relationship between material conditions and health. The thesis is that through a critical awareness of this relationship, Actions for improvements can be expected, which in their turn will build the strength to resist.

Dorn adopts a definition of health from the World Health Organization, which includes the physiological, psychological and social sides of the human being. He also focuses on the promotion of such health, that is *health careers*. He designs a series of classroom units, in which the pupils are led to investigate the relationships between environmental factors and their own health.

His theoretical model thus comprises:

A. Environmental factors (such as working conditions, housing conditions, leisure possibilities etc.);

B. Cultural production within social groups as a result of environmental factors;

C. Social behaviour as part of that cultural production: meals, uses of alcohol, sexuality etc.;

D. Effects on health and welfare.

There is no conception of Activity here. But Dorn makes it easier for us to envisage such. He wants the pupils to find out what might be bad for their health and welfare. He guides the teacher and her pupils towards the right door, Activity. Dorn does not explicity mention the use of mathematics in his design of the classroom units. It is not difficult to see the significance of its use. To demonstrate this, we can explore the possibilities of a project familiar to many primary school teachers.

Several textbooks suggest quantifying the day and night time activities of the pupils: Draw a circle. Divide it in 24 equal sectors, and number them

- from 1 to 24, 24 at top;
- Colour yesterday's activities. When did you sleep? Go to school? Work (deliver the papers, help your mother and father, etc.)?
- When did you participate in *organised* activity (scouts, music, sport etc.)?
- When were you with friends? By yourself?

Many textbooks stop there, reducing this task to the mere collection of data and some analysis of the distribution. But to do that would leave the pupils confused: what is all this about?

As regards drug education the vulnerable period is unorganised time. In the context of such education one would, in order to help pupils spot typical drug situations, ask: If drugs were present would that be in specific situations (places and persons)?

Here I shall prefer to approach a possible Activity more generally:

If you could choose, what would your activity circle be like then?

More data is collected and can be compared with the first set. Experience tells us that it is frequently necessary to introduce a third project:

What things can you do around here? Within a distance of 2 km, say.

It is often surprising how ignorant pupils can be about leisure possibilities, organised or unorganised, in the near neighbourhood. It is equally surprising to see how the adult world can exaggerate: the local sports club exists but only has room for a limited number of active members, the youth club has to reject half the applicants, and so on.

A fourth project is now suggested by the first three:

- What can be done to reduce the discrepancy between possibilities and wishes?
- What can we contribute, and what help do we need from the adult world?
- How can we approach the adult world with our justified claims?

With the last question Activity enters on the scene to its full extent as a result of the previous investigations and analyses. The main differences between this project and the Moving project and the project of Fasheh are that it seizes the possible Activity from the very beginning; its design includes such a possibility.

Project

A dormitory town outside Bergen. High, grey concrete blocks of flats. The area was declared ready for occupation as soon as the garages had been built. It was not thought scandalous until later it was discovered that they had forgotten to provide the children with leisure areas.

The teachers of three classes (age group 10) prepared a project. What can we do about the situation?

To start with, the teachers first provided a list of possible leisure activities and the pupils had to decide their preferences. Later the pupils made their own list and interviewed the rest of the pupils (ca. 400). Data collection, statistics, data analysis.

One of the teachers told this story: a pupil of his discovered that after he had written a report to be read by the authorities he could not read his own handwriting. The teacher had been trying to persuade this kid to improve his handwriting for years. For the first time ever the boy now recognised the value of his teacher's advice. Handwriting as a tool had finally become functional.

The teachers and their pupils pursued their task by approaching the various boards and councils in charge who might support their demands.

This project is the nearest I have come to initiating some Activity where mathematical knowledge is required, and thus becomes functional knowledge. The Activity emerges as the analysis and discussions are developed into Action.

One major weakness of this project, as it turned out, was that the parents interfered too much when Action came. To some extent this reduced the pupils' responsibility for their own project. The project provides an example of how ten-year-olds could take responsibility at quite a high level. The dads and mums were, however, a bit too eager to assist and reduced the opportunities for children's responsibility.

Another common problem with such projects did not arise. What would have happened if the children had had to press for their demands? If they had been met with negative attitudes? If they had been met with negative attitudes and no concrete help (i.e. a double-bind situation). Imagine:

"We could start by kicking their cars. I'll start collecting wing mirrors tonight."

The teachers themselves face a double-bind here. If they do not stand up for their pupils in a situation like the one described above the pupils will lose faith in them. On the other hand there is a frame factor as represented by the degree of tolerance of the educational system in any society. I am not underestimating the importance of the latter. In the end, however, teachers act politically when practising their profession too.

5.2.6. *Using the Micro*

The type of project reported below is common in computer education at Bergen College of Education. It is based on a principle which is closely allied to Activity theory: computing within school should reflect its real power outside school. We consider this power of the micro to handle a huge collection of data in a very short time, thus extending man's capacity to handle information. This broadens the range of possible projects in school. One important aim will be to demonstrate the functionality of the micro as a tool in this way.

Project:
CP/M operational system or the equivalent.

A sorting system such as Supersort and a word processing system such as WordStar.

(These facilities are compulsory in the secondary school standards set by the Norwegian authorities.)

Task: To organise the school sportsday
There are n pupils of age $(a_1, \ldots, a_n)$. They belong to classes $(c_1, \ldots, c_m)$. There will usually be two indexes here. They participate in events $(e_1, \ldots, e_p)$.

First some questions to be answered:

- What are the data records and which fields are needed?
- Can some efficient algorithm be constructed for the data-run according to the records and files required?

One algorithm might look like this:

1. By the use of word processing (wp) the pupils are recorded for participation according to their preferences. This is file F(1).

2. By the use of the Select- and Sort-command of the sorting system (Sort), F(1) divided into subfiles consisting of lists of participants in each event, ordered according to age (or class). This gives files *Fst*(2) where *s* varies over the events and *t* over age.

3. The order of participation is determined according to a system. Sometimes randomising is appropriate. Some sports, such as swimming, have intricate systems which involve personal records and so on. These become files *Fstu*(3) where *u* represents order of participation (such as group, division, heat).

4. As the results come in, F(3) is updated by the wp producing a merged main file of result F(4).

5. F(4) is sorted according to certain criteria. Many schools avoid sorting according to the best performances and choose to employ some deviation from mean performance or the use of a random result as a criterion. F(4) is thus transformed into *Fst*(5), which forms the final results lists.

Part of the algorithm will be to decide which groups of people are to receive which printouts. What will the staff administration of the sports day require? Part of the work with this algorithm will also be to test it on a limited number of entries, and to see whether it can be made simpler and more efficient.

5.2.7. *The Importance of the End-product*

The end product of a project calling for an Activity has to be considered in the initial stages of the project. The most common way to report a project is to put up a report on the wall or to produce a booklet to give out and read in the class. Such end-products will usually not be sufficient for the purpose of an Activity.

Consider the following cases:

A. The class produces a white paper about the need to use the school playground in the afternoons. It concludes that floodlights are required during the winter-time. The white paper is presented to the School Board.

B. There is increased driving on red at a nearby junction, causing a new danger for pupils going to school. Furthermore, it turns out that the smallest children cannot see the traffic lights when the lorries come between. The class produces systematic evidence about this. It writes to the newspaper and reports to the local police demanding increased control and a new position for the traffic lights.

C. The minimum age limit of the youth club is 16. This causes great frustration for the 12—15 year-olds who have few other leisure opportunities. Documentation is collected and a meeting is held in which the social workers responsible for the club participate.

The end product of a project also belongs to media education: how are important information, experience and demands reported and documented?

How can the use of means of visualisation such as graphs and other pictorial models be developed for those purposes? The end product will naturally be dependent on the level of ambition:

- Is the purpose conscientisation?
- Is it to influence other people?
- Is it for the purpose of change?
- If the goal is to influence people in order to change the present situation, what will the follow up be if negative attitudes are met?
- What if the School Board say no to flood lights, not necessarily because of the costs, but because they argue that the demand is unjustified?
- What if the social workers say that there is no way at all that 12—15 year-olds can get to join the club, and that they cannot see any need for them to do so anyway?
- What are the considerations about negative attitudes and the eventual follow up?
- What are the considerations about our own efforts to obtain the changes we want?

5.2.8. *Which Mathematics?*

So far I have paid almost no attention to the exploitation of the various thinking-tools used in the various projects. When reviewing the projects, it is striking to see how familiar tools have been used in new ways and how the need for other tools has come up.

In Bergen we have an in-service course in mathematics which counts as a 1/2 year unit of graduate education. Ten years ago we started with the elementary theory of functions and vector geometry in the traditional way. Not very much is left of the original curriculum today.

The theory of functions has gradually acquired a stronger component of discrete mathematics. This development has intensified as we have directed our efforts towards the identification of powerful thinking-tools in fuction theory. "Powerful" here relates to the didactical context, that is, with a regard for the pupils' needs as well.

Several of the projects reported throughout this book were carried out by the teachers on these courses. Others were done by students on our general teacher-training courses.

The transformation of the curriculum is best demonstrated in the theory of functions. As we studied the variety of growth of empirical data (wages, employment, youth club capacity) we experienced the need for numerical methods. Examples of thinking-tools in frequent use are:

1. Percentages and ratios as a means of comparision;
2. Linear functions of the form $y = ax + b$, with particular regard to a as a growth factor;
3. The approximation of a continuous function by a finite set of linear functions over closed intervals — an investigation of the gradients of these linear functions to find out about any patterns of growth;
4. The exponential function as a model, and the link between its doubling properties and its growth rate as measured by percentage — this model is an excellent tool even among the 12-year olds as the pupils can examine empirical growth patterns by looking for their doubling properties; the percentage increase can also easily be found using a calculator as well;
5. the concept of average growth over a closed interval;
6. calculations of area under a curve to obtain an expression for the total number (population, consumption); numerical methods for this;
7. various methods of graphical representation;
8. optimisation problems in cases where some materials are given and a product is wanted;
9. simple use of Basic programming to extend the possible range of analysis.

The above represents some of the uses of thinking-tools which might be required for our purposes. I should not be surprised if discrete mathematics, numerical and statistical methods combined with the development of graphics would be increasingly important in the future for the purpose of imbedding outside mathematics into mathematics theory in the context of teacher-training. This is also in line with the opportunities offered by the micros.

5.2.9. *The Dialectics between Inside and Outside Mathematics*

The previous section points out a serious deficit of present mathematics education. This book is no exception. The research on the dialectics between inside and outside mathematics at the levels of education (from primary school to teacher education) is incomplete and poor.

One of the mathematicians in Europe who has been most concerned with this dialectics is the Dane, Mogens Niss. He constantly stresses the need to take mathematics seriously and not reduce the subject to a set of trivialities in the educational context. He states the goals for a mathematics education in a way which recognises its political components:

> — to enable students to realize, understand, judge, utilize and sometimes also perform, the application of mathematics in society, in particular to situations which are of significance to their private, social and professional lives.
>
> Niss, 1983, p. 248.

Following this Niss proposes two goals for mathematics education:

> a. Students should *acquire understanding of those factors* within mathematics (such as ideas, concepts, edifices of theory, methods, etc.), as well as those outside mathematics, which are of importance to the applicability of mathematics, its potentials and limitations;
>
> b. students should themselves acquire experiences with applying, independently and in a non-mechanical manner, *mathematics* in treating extra-mathematical situations.
>
> Ibid., p. 248

In order to achieve such goals Niss develops his concept of an *integrated* course, and sees the need for two different courses:

> 1. integrated courses where the main interest lies with the non-mathematical subjects and where mathematics is a service introduced only when and only to the extent that it serves these main interests;
>
> 2. integrated courses intending to serve also purposes paying particular regard to aspects of mathematics.
>
> Ibid., p. 249

This relates to the conceptions of "wide" and "narrow" studies of the thinking-tools of Activity theory. Clearly, the two courses such as Niss describes have to be planned in relation to each other.

Jahnke (1978) and Otte (1984a) join in here as well, supporting the view that mathematics as theoretical knowledge cannot be incorporated with its applications. Otte (ibid.) emphasises applications or examples of a theory as metaphors of the theory. It is such views which have made me advocate a curriculum based on a study of the mathematical structures connected with the thinking-tools of the subject and their applications outside mathematics in order to develop an appropriate metaconcept of mathematics among pupils.

The two processes, the narrow and the wide, connect dialectically, although they are separate in form. It is this dialectic which raises the problem which is yet to be solved. It has probably to do with a failure of the mathematics educator to grasp the historical dimension in curriculum planning: What are the historical reasons for today's curriculum, and how does this fit in with today's demands on the members of society?

The kind of dialectics which I call for is exemplified in the previous paragraph concerning the elementary theory of functions: after stating by building the theory of functions mainly on the basis of continuous functions, we structured the mathematical knowledge more and more on the basis of discrete functions.

My contribution throughout this book has been to specify some necessary conditions for such a dialectics, merely demonstrating fragments of a curriculum for Niss's course 2, quoted above. The demand for a dialectics as suggested here indirectly implies a criticism of courses based solely on mathematical modelling and approaches where curriculum makers, such as textbook writers, consultants and teachers are constantly searching for "interesting" and "practical" applications of theoretical concepts.

It is hard to see how such approaches can relate to the theoretical aspects of mathematics appropriately and thus build a functional metaconcept of mathematics. There is still a long way to go before we have gained experience and accumulated knowledge about this dialectic. Mathematics educators need not be pessimistic about the future. We have a great deal of experience in building curricula which consists of purely theoretical knowledge. We have experience of mathematical modelling. There should be some way of combining various approaches, taking the best out of each in order to meet our pupils where they are.

NOTES

INTRODUCTION

[1] Examples of works about the experiences and reactions of the pupils in relation to school are Goodman (1971); Haskins (1967); Henry (1963, 1970, 1976); Holt (1969, 1970); Kohl (1972); Kozol (1968); O'Neill (1970) and Silberman (1970).

CHAPTER 1

§1.1.

[1] See Note 3 §2.1.

[2] Research about Folk Mathematics as carried out in Bergen is reported in Hermansen (1980) and Mellin-Olsen (1979, 1980).

[3] I am indebted to Einar Jahr for the notion of *intellectual material.* Jahr is a mathematics educator, but he used the expression when reviewing classical music.

[4] Len S. Vygotsky (1896—1934) was born the very same year as Piaget. There are several points of contact between their theories. The main contradiction is on the question of the role of language for the development of intelligence. Vygotsky had already published in 1934 a critique of Piaget's conceptualisation of ego-centric speech in the child, a critique which Piaget first discovered and responded to 25 years later. This exchange of theoretical views is reported in Vygotsky (1962). Vygotsky was educated as a specialist in literature. A more detailed bibliography is given by the editors of Vygotsky (1979).

§1.2.

[1] The quotation is taken from Couvent (1979). The Persepolis Symposium was about literacy and was held from 3—8 September 1975. It was administered by UNESCO.

[2] Indeed, I am not alone in having problems in discovering the significance of Vygotsky's use of "activity". Wertsch (1981) makes the same point. Even in recent elaborations Vygotsky's theory is reduced to a theory about "inner language" or about language in its cultural and historical context. Such a conception is far too limited if we are to go along with Vygotsky's theory. A recent example of such a crippling reference to the theory is Saxe and Posner (1983). Having traced American and British research connected with the theory, I find Elsasser and John-Steiner (1977) as one of the few who catch the spirit of Vygotsky's thinking. See §4.2.

[3] I have a problem of notation, and I have chosen what I myself see as an unsatisfactory solution. I choose to use capital letter: Activity, to stress the specific use of the term. The origin of the word in Soviet psychology is probably Marx's "Tatigkeit" (used by Marx already in the theses on Feuerbach), which was translated into "dejatelnost"

in Soviet (Enerstvedt 1982). Wertsch (1981) uses the spelling "deyatel'nost" here. The problem of spelling occurs also with "Vygotsky/Wygotsky/Vigotsky and Leontjev/Leont'ev.

In Scandinavian languages the word "virksomhet" is used for Activity. "virke" can be traced to the Indo-German word "werg", which becomes "Werk" in German and "work" in English. Enerstvedt (op. cit.) gives a thorough analysis of the roots of these words and other basic terms used in Activity theory.

[4] Enerstvedt refers to Lomow (1980) and Leont'ev (1980) about this discussion.

[5] My use of "praxis" is the same as Mao's. It refers to "practice + ideology/theory". By referring to "praxis" one is thus also referring to the intentions/policy connected to the behaviour.

[6] *ontogeny*: the origin and development of the individual being. *phylogeny*: the history and development of the species. It is the idea of both Piaget and Vygotsky that by tracing phylogeny one can obtain better insight in ontogeny. For Vygotsky this implies as investigation into man's historical development of tool use, including communicative tools. Piaget stresses the development of mother structures of human intelligence: topological, algebraical and order structures, and their corresponding development in ontogeny (Beth and Piaget 1966). Freudenthal (1983) repeatedly explores the phylogeny of mathematics for its possible influence on ontogeny.

[7] All Freire's writings are expositions on the theme of functional literacy (Freire 1970, 1970a, 1972, 1975, 1978, 1981). Analysis of "functionalism" in this context is also given by Baratz and Baratz (1970); Bowles (1971); Cardenal and Miller (1981); Elsasser and John-Steiner (1977); Fisman (1979); Levine (1982) and Mackie (1980).

CHAPTER 2

§2.1.

[1] In Scandinavia, Holmberg and Malmgren (1979) have developed a methodology for language education on the principle of children's own needs for uses of language. This issue is further discussed in §3.4. where a critique of Chomsky is made, and in §4.2. where psychoanalysis is connected with oppression of language use.

[2] Dawe refers to Cummins (1978, 1979). The problem is further discussed in §§3.3. and 3.4. Further support for Cummin's hypothesis and the stress on the necessity of a functional L1 language is provided by BBC (1981); Fisman (1979); Hakuta and Cancino (1977); Teitelbaum and Hiller (1977) and Smitherman (1981). The latter two may be of particular interest as they report court decisions from trials where schools have been charged for neglecting the treatment of the L1 language among black children.

[3] The Madison Project was revolutionary at its time in the way it challenged the pupils to employ their own thinking and to exploit it at the same time as this thinking was explored and discussed in the social context of the classroom (Davis 1967, 1980). Later Davis founded and edited a special journal on the topic: *The Journal of Mathematical Behavior*. This journal reports a series of studies on children's inventive mathematical constructs. See also §2.3.3.

CHAPTER 3

§3.1.

[1] I can only offer some of the definitions we can meet in the complicated landscape of visualisation:
spatial ability: the ability to formulate mental images and to manipulate these images in mind.
imagery: the occurrence of mental activity corresponding to the perception of an object, but when the object is not presented to the sense organs (Lean and Clements, 1981).

Liber (1981) has, according to Clements (1983) the potential for unifying the concepts involved:
spatial products refer to external products that represent space in some way. Any kind of external representation of space is a spatial product.
spatial thinking refers to thinking that concerns or makes use of spatial representations in some way; it is knowledge that individuals have access to, can reflect upon, or can manipulate.
spatial storage refers to any information about space which, although contained "in the head", is such that the individual is not cognisant of it.

[2] Fennema refers to Fennema (1975) and Fennema and Sherman (1978).

§3.2.

[1] Edward Sapir (1884—1939) began his career in charge of the anthropological research at the Canadian National Museum before moving to the University of Chicago (1925) and Yale (1931). Whorf (1897—1941) has been described as an outstanding example of the brilliant amateur in scholarly work (Sampson 1980). He had a degree in chemical engineering and had a successful career as a fire-prevention inspector with an insurance company in Connecticut. He worked for this company until his death. When Sapir moved to Yale, Whorf became his regular collaborator, working mainly on the Hopi.

[2] Sampson refers to Hjelmslev (1963), Lamb (1966) and P. A. Reich (1970).

[3] I shall return to some important research on colour discrimination in §3.4.5.

§3.3.

[1] About China at the time of the cultural revolution, see Chang-Fu Hu (1974) and Manger et al. (1974). About later changes in Chinese education, see Løfsted (1980). On mathematics education in the People's Republic of China in particular see Swetz (1978) and Swetz and Ying-King Yu (1979).

[2] Reproduction of society is a complicated field of sociological research, which has been extensively explored over the last few decades. I shall return to the topic in Chapter 5, but so far I can refer to Althusser (1972), Bernstein (1975), Bourdieu and Passeron (1977) and Masuch (1974), as representatives for British (Bernstein), French (Althusser, Bourdieu and Passeron) and German (Masuch) sociology. Representative collections of papers in the field have been provided by the following as editors: Brown (1973), Cosin (1972), Gleason (1977) and Karabel and Halsey (1977). See also §5.1.1.

[3] See Hua (1983). The responses to his lecture were made by Bernstein (1983) and Gyakye Jackson (1983).

§3.4.

[1] Some of the recent research about the importance of contextual interpretations for language acquisition is reported by Chafe (1970); Elsasser and John-Steiner (1977); Feldman (1977); Hymes (1980); Leonard (1976); Labov (1972a); McNeill (1970) and Mishler et al. (1979).

[2] Clements refers to Casey (1978) and Newman (1977).

[3] Erlwanger (1973) gives a powerful case study about this. Ginsburg (1977) also provides striking evidence about the contradiction between children's arithmetical inventiveness and the control-factors. See §2.3.3.

CHAPTER 4

§4.1.

[1] George Herbert Mead (1863—1931) was four years younger than John Dewey, who was his best friend during his life time. Mead joined Dewey at the University of Chicago in 1894. He never published any books: his writings have been compiled by his students and followers. The most reputed book, *Mind, Self and Society* was put together from notes made by his students. Mead's interactionist theory has achieved immense influence in the research territory between sociology and psychology during the last decades, and can be said to have laid the foundation for a particular theoretical tradition: interactionism.

[2] In Norway and probably many other countries as well there is an increasing concern about the autonomy of youth cultures in relation to the drug problem.

An increased social autonomy among young people can lead to new forms of social knowledge, as previous knowledge disappears. We are thinking here of the social knowledge which is necessary for the individual to cope with important situations. A typical example of this is the knowledge about how to behave in "desk"-situations such as in a bank, at the employment office and so forth. In drug education it is an important task to teach young people how to cope in situations where there is a risk of drug use by developing a functional language for these.

Mathematics educators have not, as far as I know, discussed the implication of this development for their subject. We have hardly discussed the social significance of the "back to basics" trend in mathematics experienced in many countries.

[3] Taking up the notion of *ideology* is a risky thing, as few fields of research are so diverse and confusing as this. But as I have claimed that Activity is political in nature, I can hardly avoid an analysis of the concept.

Recent developments in ideology theory seem to clarify some of the old confusions: conceptions such as *false consciousness* and *false ideology* are sorted out, and it is stressed that people, not only institutions, represent ideology.

I have profited from Adlam et al. (1977); Adlam and Salfield (1978); Covard and Ellis (1977); Hall (1977, 1978) on this: they all reinstate the individual as a representative of ideology.

[4] The phenomenon hinted here illustrates the principle of exponential growth in communication theory (Bateson 1973). The situation is as follows: a communication structure C develops behaviour B and B contains unwanted elements B′. In order to

eliminate B′ an insufficient analysis of the communication leads to a more efficient use of elements in C which does not disturb its structure. The result will be an increase in B′. If this again results in a more efficient use of C, the result will be an exponential growth in B′. Bateson's (ibid.) case is pollution and an inappropriate analysis which implies that the remedial programmes for reducing the analysis leads to new pollution. Bateson's major point here, which is mine as well, is the necessity to treat human communication systems as open systems. This implies that if pollution occurs within the system one should look for repair tools outside it.

§4.2.

[1] Freud (1856—1939) needs hardly any biographical introduction. Of the grand theorists I build on in this book he is the oldest, belonging to the same generation as G. H. Mead. Later in this section we shall meet the Soviet semiologist V. N. Volosinov (b. 1895), who wrote a brilliant critique of Freud's work. This critique still has much to say to the modern cavalry of sociolinguists. As he is a contemporary of Vygotsky it is interesting to see how close the two are in their thinking: both work according to the thesis that language is the bearer of social meaning. According to his biographers, Bruss et al. (1976), Volosinov died in the cold times of the mid-1930s, as nothing has been heard from him since.

From now on we shall also hear more about Paulo Freire. He was born in 1921 in the town of Recife on the east coast of Brazil. He was a teacher devoted to Catholicism. Having problems in coping with the life style of the members of his Christian community, he and his wife settled in the workers' ghetto in Recife. "God took me to the people and the people took me to Marx" has become one of Freire's classic sayings.

He soon developed his programme for literacy, based on methods which I shall return to in more detail in Chapter 5. As the success of his programme spread, Freire's influence increased correspondingly, and he ended up as Secretary of Education in his country. He also held the chair in education at the University of Recife. In 1964 the liberal government in Brazil was deposed by the military in a coup d'état, and Freire went into exile after a short spell in prison.

Since 1970 he has held an appointment as a special consultant to the office of Education at the World Council of Churches in Geneva. The reader's attention is drawn to his latest book, *Pedagogy in Progress: The Letters to Guinea-Bissau*, Seabury Press, New York, 1978. Here he explores and clarifies his previous positions as a result of his latest innovative work. For a more detailed biography see Mackie (1980). Giroux (1981) also gives a theoretical evaluation of Freire's works. See Note 7, §1.2.

[2] Althusser (1970) and Althusser and Balibar (1970) stress the importance of reading Freud historically and structurally. Levi-Strauss (1958) defined the subconscious, with reference to Freud, as the location for the symbolic function, which represented the possible messages within the kinship systems studied by structural anthropology. Yet it is primarily Lacan who opened the way for a renaissance of Freud by building his psychoanalysis on the foundations of language: the subconscious is structured as a language.

Wilden (1968, 1972) contributes a deep analysis of French structuralism and the works of Gregory Bateson from the perspective of communication and exchange theory.

Freud's works have been a tempting field for exploration for radicals of all stripes since the time they appeared. Fromm, Reich and Marcuse are names which easily come to mind. As Freud's conceptual constructs relate closely to those of ideological theory, they still repeatedly come up in the context of political theory and, as in my case, in educational theory.

[3] The study of classroom communication is worth a book in itself and, indeed, several have been written. I can here only emphasise a major point from the perspective of Activity theory: the need for the teacher to avoid too much routinisation of classroom registers. The teacher can hardly avoid using such registers ("Good morning, everyone"), but should bear in mind that they produce the possibility of the pupils communicating knowledge which may be of importance to *them*. Such careful control of classroom registers is a skill which it is to be hoped can be trained in teacher education. It is, however, also related to frame factors: if these are rigid we may expect the control of registers to be correspondingly rigid. Analysis and discussion of classroom communication is reported by Bauersfeld (1980); Bishop and Goffree (1984); Cazden et al. (1972); Edwards (1976); Edwards and Furlong (1978); Elsasser and John-Steiner (1977); Kemme (1981) and Stubbs (1976, 1976a).

§4.3.

[1] Gregory Bateson was born in 1904 in Grantchester near Cambridge. He graduated with an M.A. in anthropology in 1930 at St. John's College, Cambridge, and was a fellow at the same college until 1937. In this period he wrote his first book *Naven*. In 1940 he emigrated to the United States. During the war he worked at the Office of Strategic Services of the US Government.

Besides his work on a communication theory of schizophrenia, his research interests covered fields such as the theology of Alcoholics Anonymous (Bateson 1973), dolphin communication and ecology. At his death in 1981 he was a Visiting Professor at the University of California at Santa Cruz. A detailed biography is given in Brockman (ed.) (1977).

[2] It is especially the British school of psychiatry built by Ronald D. Laing which has exploited Bateson's theory. The works of Laing (1968), Laing and Cooper (1970) and Cooper (1971) are intriguing to read for the educationist who can translate "hospital" into "school". However, one has to bear in mind that Laing and his colloborators never considered the material frames of families. Their interest was solely in terms of internal communication, not about the circumstances in which this communication took place.

[3] The dialectical nature of Bateson's thinking, making it a nondeterministic theory, is his conception of the human communication system as an *open* system. According to this, metalearning cannot be deterministic as regards learning. See Note 4 §4.1.

CHAPTER 5

§5.1.

[1] Over the last few years a *sociology of resistance* has developed. See Corrigan (1979); Giroux (1981, 1983); Hall and Jefferson (1980), Hudson (1983); Jenkins (1983) and Willis (1977, 1981).

§5.2.

[1]Nick Dorn has worked for the *Health Careers Teaching Project*, 3 Blackburn Road, London NW 6 IXA. Together with the Dane Bente Nortorf he wrote *Health Careers. A Thirteen-unit teacher's manual for use with school-leavers and further education students* for the Project.

REFERENCES

Adlam, D.: 1977, 'Psychology, Ideology and the Human Subject,' *Ideology & Consciousness* **1**.

Adlam, D. and Salfield, A.: 1978, A Matter of Language,' *Ideology & Consciousness* **3**.

Althusser, L.: 1970,'Freud and Lacan,' in P. Madsen (ed.), *Strukturalisme*, Rhodos, Copenhagen.

Althusser, L.: 1972, 'Ideology and Ideology State Apparatus,' in B. R. Cosin (ed.), *Education: Structure and Society*, Penguin Books and The Open University, Harmondsworth.

Althusser, L. and Balibar, E.: 1970, *Att Lasa Kapitalet I och II*, Cavefors, Stockholm.

Altvater, E.: 1974, 'Kvalifikation og erhvervsuddannelse,' *Nordiska Sommeruniversitetet*.

Austin, J. L. and Howson, A. G.: 1979, 'Language and Mathematical Education,' *Educational Studies in Mathematics* **10**.

BBC: 1981, *Multi-Cultural Education. Views from the Classroom*, BBC, London.

Baratz, S. B. and Baratz, J. C.: 1970, 'Early Childhood Intervention: The Social Science Base of Institutional Racism,' *Harvard Educational Review* **40**.

Bateson, G.: 1973, *Steps to an Ecology of Mind*, Paladin Books, Herts.

Bauersfeld, H.: 1979, 'Research Related to the Mathematical Learning Process,' in *New Trends in Mathematics Teaching* IV, UNESCO, Paris.

Bauersfeld, H.: 1980, 'Hidden Dimensions in the So-called Reality of a Mathematics Classroom,' *Educational Studies in Mathematics* **11**.

Berger, P.: 1971, 'Identity as a Problem in the Sociology of Knowledge,' in *School and Society. A Sociological Reader*, Routledge and Kegan Paul and The Open University Press, London.

Berlin, B. and Kay, P.: 1969, *Basic Color Terms*, University of California Press, Berkeley.

Bernstein, B.: 1975, *Class, Codes and Control I-III*, Routledge and Kegan Paul, London.

Bernstein, D.: 1983, 'Reaction to Hua Loo-Keng's Plenary Lecture,' in *Proceedings of the Fourth International Congress on Mathematics Education*, Birkhauser, Boston.

Beth, E. W. and Piaget, J.: 1966, *Mathematical Epistemology and Psychology*, D. Reidel, Dordrecht.

Bishop, A. J.: 1979, 'Visualising and Mathematics in a Pre-Technological Culture,' *Educational Studies in Mathematics* **10**.

Bishop, A. J.: 1980, 'Spatial Abilities and Mathematics Education. A Review,' *Educational Studies in Mathematics* **11**.

Bishop, A. J. and Goffree, F.: 1985, 'Classroom Organization and Dynamics,' in B. Christiansen, A. G. Howson and M. Otte (eds.), *Perspectives on Mathematics Education*, D. Reidel, Dordrecht.

Bishop, A. J. and Whitfield, E.: 1972, *Situations in Teaching*, McGraw Hill, London.

Blumer, H.: 1971, 'Sociological Implications of the Thought of George Herbert Mead,' in *School and Society. A Sociological Reader*, Routledge and Kegan Paul and The Open University Press, London.

Booth, L. R.: 1981, 'Child Methods in Secondary Mathematics,' *Educational Studies in Mathematics* **12**.

Bourdieu, P.: 1977, *Outline of a Theory of Practice*, Cambridge University Press, London.

Bourdieu, P. and Passeron, J. C.: 1977, *Reproduction in Education*, Sage Publ., London.

Bowles, S.: 1971, 'Cuban Education and the Revolutionary Ideology,' *Harvard Educational Review* **41**.

Brockman, J. (ed.): 1977, *About Bateson*, E. P. Dutton, New York.

Brousseau, G.: 1984, 'The Crucial Role of the Didactical Contract in the Analysis and Construction of Situations in Teaching and Learning Mathematics,' in H. G. Steiner (ed.), *Theory of Mathematics Instruction*, Occasional Paper, IDM, Bielefeld.

Brown, J. S. and Burton, R. R.: 1978, 'Diagnostic Models for Procedural Bugs in Basic Mathematical Skills,' *Cognitive Science* **2**.

Brown, R. (ed.): 1973, *Knowledge, Education and Cultural Change*, Tavistock Publications, London.

Bruner, J. S. et al. (eds.): 1966, *Studies in Cognitive Growth*, Wiley and Son, New York.

Bruner, J. S.: 1972, *The Relevance of Education*, Allen & Unwin, New York.

Bruner, J. S.: 1974, *Beyond the Information Given*, Allen & Unwin, New York.

Bruss, N. and Titunik, I. R.: 1976, Preface in V. M. Volosinov, *Freudianism: A Marxist Critique*, Academic Press, New York.

Burkhardt, D.: 1983, 'Real Problemsolving,' in *Proceedings of the Fourth International Congress on Mathematics Education*, Birkhauser, Boston.

Cardenal, F. and Miller, V.: 1981, Nicaragua 1980: 'The Battle of the ABC,' *Harvard Educational Review* **5**.

Casey, D. D.: 1978, 'Failing Students: A Strategy of Error Analysis,' in P. Costello (ed.), *Aspects of Motivation*, Math. Ass. of Victoria, Melbourne.

Cazden, I. B. (ed.): 1972, *Functions of Language in the Classroom*, Teachers College Press, New York.

Chafe, W.: 1970, *Meaning and the Structure of Language*, University of Chicago Press, Chicago.

Chang-Fu Hu (ed.): 1974, *Chinese Education under Communism*, Teachers College Press, Columbia University Press, New York.

Chomsky, N.: 1959, 'Review of Skinner's Verbal Behavior,' *Language* **35**.

Chomsky, N.: 1972, *Language and Mind*, Harcourt Brace Johanovich, New York.

Christiansen, B.: 1984, 'Selvvirksomhed og erfaringer i matematikundervisningen,' I T. B. Jensen og H. Vejleskov (ed.) *Fag og erfaring*, Munksgaard, Copenhagen.

Christiansen, B.: 1984a, 'The Theoretical Background of the Project: Basic Components of Mathematics Education for Teachers (BACOMET),' in H. G. Steiner (ed.) *Theory of Mathematics Instruction*, Occasional Paper 54, IDM Bielefeld.

Christiansen, B. and Walther, G.: 1985, 'Task and Activity,' in B. Christiansen, A. G. Howson and M. Otte (eds.) *Perspectives on Mathematics Education*, D. Reidel, Dordrecht.

Cicirelli, V. G.: 1970, 'A Report to the Report Analysis,' *Harvard Educational Review* **40**.

Clements, M. A.: 1980, 'Analyzing Children's Errors on Written Mathematical Tasks,' *Educational Studies in Mathematics* **11**.

Clements, M. A.: 1982, 'Visual Imagery and School Mathematics,' *For the Learning of Mathematics* **2**.

Clements, M. A.: 1983, 'Visual Imagery and School Mathematics', *For the Learning of Mathematics* **3**.

Cole, M.: 1981, Preface, in J. Wertsch (ed.) *The Concept of Ability in Soviet Psychology*, M. E. Sharpe, New York.

Coleman, J.: 1968, 'The Concept of Equality of Educational Opportunity,' *Harvard Educational Review* **38**.

Cooper, D.: 1971, *Psykiatri og anti-psykiatri*, Gyldendal, Oslo.

Cooper, D. and Clancy, D.: 1982, *Oh! Pascal!*, W. W. Norton, New York.

Corrigan, P.: 1979, *Schooling the Smash Street Kids*, The Macmillan Press, London.

Cosin, B. R. (ed.): 1972, *Education: Structure and Society*, Penguin Books for the Open University, Harmondsworth.

Couvent, R.: 1979, *The Evaluation of Literacy Programs*, The Unesco Press, Paris.

Coward, R. and Ellis, J.: 1977, *Language and Materialisms: Developments in Semiology and the Theory of the Subject*, Routledge and Kegan Paul, London.

Cummins, J.: 1978, 'Educational Implications of Mother Tongue Maintenance in Minority Language Groups,' *Canadian Modern Language Review* **34**.

Cummins, J.: 1979, 'Linguistic Interdependence and the Educational Development of Bilingual Children,' *Review of Educational Research* **49**.

Davis, R.: 1967, *Explorations in Mathematics: A Text for Teachers*, Addison Wesley, Palo Alto.

Davis, R.: 1980, *Discovery in Mathematics: A Text for Teachers*, Cuisinaire, New York.

Davis R.: 1983, 'Complex Mathematical Structures,' in H. Ginsburg (ed.), *The Development of Mathematical Thinking*, Academic Press, New York.

Davis, R. and McKnight, C.: 1979, 'Cognitive Processes in Learning Algebra,' *Journal of Children's Mathematical Behavour* **2**.

Davydov, V. V. and Markova, A. K.: 1982—83, 'A Concept of Educational Activity for Schoolchildren,' *Soviet Psychology* 21.

Dawe, L.: 1983, 'Bilingualism and Mathematical Reasoning in English as a Second Language,' *Educational Studies in Mathematics* **14**.

Dienes, Z. P.: 1964, *Building up Mathematics*, Hutchinson, London.

Dienes, Z. P.: 1967, *The Power of Mathematics*, Hutchinson, London.

Douglas, J. D.: 1973, *Understanding Everyday Life*, Routledge and Kegan Paul, London.

Dörfler, W. and McLone, R.: 1985, 'Mathematics as a School Subject,' in B. Christiansen, A. H. Howson and M. Otte (eds.), *Perspectives on Mathematics Education*, D. Reidel, Dordrecht.

Edelman, M.: 1975, 'The Political Language of the Helping Professions,' *Politics of Society* **4**.

Edwards, A. D.: 1976, *Language in Culture and Class*, Heinemann, London.

Edwards, A. D. and Furlong, J. W.: 1978, *The Language of Teaching: Meaning in Classroom Interaction*, Heinemann, London.

Elsasser, N. and John-Steiner, V. P.: 1977, 'An Interactionist Approach to Advancing Literacy,' *Harvard Educational Review* **47**.

Enerstvedt, R.: 1982, *Mennesket som virksomhet*, Tiden, Oslo.

Erlwanger, S. H.: 1973, 'Benny's Conception of Rules and Answers in IPI Mathematics,' *Journal of Children's Mathematical Behavior* **3**.

Fasheh, M.: 1982, 'Mathematics, Culture and Authority,' *For the Learning of Mathematics* **3**.

Feldman, C. F.: 1977, 'Two Functions of Language,' *Harvard Educational Review* **47**.

Fennema, E.: 1975, 'Spatial Ability, Mathematics and Sexes,' in E. Fennema (ed.), *Mathematics Learning: What Research Says About Sex Differences*, ERIC, The Ohio State University Columbus.

Fennema, E. et al.: 1979, 'Women and Girls in Mathematics. Equality in Mathematics Education,' *Educational Studies in Mathematics* **10**.

Fennema, E. and Sherman, J. A.: 1978, 'Sex-Related Differences in Mathematics Achievement and Related Factors: A Further Study,' *Journal for Research in Mathematics Education* **9**.

Firth, J. R.: 1957, *Papers in Linguistics 1934–1951*, Oxford University Press, London.

Fisman, J. A.: 1979, 'Standard versus Dialect in Bilingual Education,' in H. T. Trueba and Barnett-Mizrahi C. (eds.) *Bilingual Multicultural Education and the Professional: From Theory to Practice*, Newbury.

Freire, P.: 1970, 'The Adult Literacy Process as Cultural Action for Freedom, *Harvard Educational Review* **40**.

Freire, P.: 1970a, 'Cultural Action and Conscientization,' *Harvard Educational Review* **40**.

Freire, P.: 1972, *Pedagogy for the Oppressed*, Penguin, Harmondsworth.

Freire, P.: 1975, *Education for Critical Consciousness*, Sheed and Ward, London.

Freire, P.: 1982, 'The People Speak their Word: Learning to Read and Write in Sao Tome and Principe,' *Harvard Educational Review* **51**.

Freud, S.: 1920, *Det Ubevidste*, Martins Forlag, Copenhagen.

Freudenthal, H.: 1978, *Weeding and Sowing*, D. Reidel, Dordrecht.

Freudenthal, H.: 1981, 'Major Problems of Mathematics Education,' *Educational Studies in Mathematics* **12**.

Freudenthal, H.: 1983, *Didactical Phenomenology of Mathematical Structures*, D. Reidel, Dordrecht.

Gay, J. and Cole, M.: 1967, *The New Mathematics and an Old Culture, A Study of Learning among the Kpelle of Liberia*, Holt, Rinehart and Winston, New York.

Gerdes, P.: 1985, 'Conditions and Strategies for Emancipatory Mathematics in Underdeveloped Countries,' *For the Learning of Mathematics* **5**.

Giddens, A.: 1976, *New Rules of Sociological Method: A Positive Critique of Interpretative Sociologies*, Hutchinson, London.

Giddens, A.: 1977, *Studies in Social and Political Theory*, Hutchinson, London.

Giddens, A.: 1979, *Central Problems in Social Theory*, Macmillan, London.

Giddens, A.: 1983, 'Four Theses on Ideology,' *Canadian Journal of Political and Social Theory* **7**.

Ginsburg, H.: 1977, *Children's Arithmetic: The Learning Process*, van Nostrand, New York.

Ginsburg, H. (ed.): 1983, *The Development of Mathematical Thinking*, The Academic Press, New York.

Ginsburg, H. and Kossan, N.: 1981, 'The Clinical Interview: Validity, Reliability and Diagnosis,' *For the Learning of Mathematics* **2**.

Giroux, H.: 1981, *Ideology, Culture & The Process of Schooling*, The Falmer Press, London.

Giroux, H.: 1981a, 'Hegemony Resistance and the Paradox of Educational Reform,' *Interchange* **12**.

Giroux, H.: 1983, *Theory & Resistance in Education*, Heinemann, London.

Gjessing, H.: 1974, 'Om sanering av specialundervisning og om alternative tilbud,' *Nord. Tidsskrift for spec. ped.* **4**.

Gleason, D. (ed.): 1977, *Identity and Structure, Issues in the Sociology of Knowledge*, Nafferation Books, London.

Goffree, F.: 1982–83, *Wiskunde & Didactiek 1.2. deel*, Wolters Noordhoff, Groningen.

Goodman, P.: 1971, *Compulsory Miseducation*, Penguin, Harmondsworth.

Gramsci, A.: 1971, *Selection from Prison Notebooks*, International Publishers, New York.

Gregg, L. W.: 1974, *Knowledge and Cognition*, Erlbaum, Hillsdale, New Jersey.

Gyakye Jackson, J. S.: 1983, 'Reactions to Hua Loo-Keng's Plenary Lecture,' in *Proceedings of the Fourth International Congress on Mathematics Education*, Birkhauser, Boston.

Hakuta, K. and Cancino, H.: 1977, 'Trends in Second-Language Acquisition Research,' *Harvard Educational Review* **47**.

Hall, S.: 1977, 'The Hinterland of Science: Ideology and the "Sociology of Knowledge",' in *On Ideology*, Centre for Contemporary Cultural Studies, Hutchinson, London.

Hall, S.: 1978, 'Some Problems with the Ideology/Subject Couplet,' *Ideology & Consciousness 3.*

Hall, S. and Jefferson, T. (eds.): 1980, *Resistance through Rituals*, Hutchinson, London.

Halliday, M.: 1978, *Language as a Social Semiotic*, Arnolds, London.

Halliday, M.: 1980, 'An Interpretation of the Functional Relationship between Language and Social Action,' in A. K. Pugh (ed.) *Language and language Use*, Heinemann and The Open University, London.

Hargreaves, D. H.: 1975, *Interpersonal Relations and Education*, Routledge and Kegan Paul, London.

Harris, P.: 1980, *Measurement in Tribal Aboriginal Communities*, Australian Council for Educational Research.

Hart, K. (ed.): 1981, *Children's Understanding of Mathematics, 11–16*, J. Murray, London.

Haskins, J.: 1969, *Diary of a Harlem Schoolteacher*, Grove Press, New York.

Heege, H. ter: 1978, 'Testing the Maturity for Learning the Algorithm for Multiplication,' *Educational Studies in Mathematics* **9**.

Heege, H. ter: 1983, 'The Multiplication Algorithm: An Integrated Approach,' *For the Learning of Mathematics* **3**.

Henry, J.: 1963, *Culture Against Man*, Random House, New York.

Henry, J.: 1970, *Essays on Education*, Penguin, Harmondsworth.

Henry, J.: 1976, 'Docility — or Giving the Teacher what She Wants,' in I. Beck (ed.), *Worlds Apart, Readings for a Sociology of Education*, Collier Macmillan, London.

Hermansen, R. (ed): 1980, *Matematikk: Til bruk i andre fag*, Caspar, Rådal.

Hjelmslev, L.: 1963, *Language: An Introduction*, University of Wisconsin Press, Madison.

Hoem, A.: 1972, *Makt og kunnskap*, Inst. for samfunnsvitenskap, Oslo.

Holmberg, O och Malmgren, L.: 1979, *Språk, litteratur och projektundervisning*, Liber, Stockholm.

Holt, J.: 1969, *How Children Fail*, Penguin, Harmondsworth.

Holt, J.: 1970, *How Children Learn*, Penguin, Harmondsworth.

Hoskyns, A.: 1977, An Experiment in the Teaching of Physics, in G. Whitty and M. Young (eds.) *Explorations in the Politics of School Knowledge*, Nafferton Books, London.

Hoyles, C.: 1981, 'Mathematics Learning — What the Pupils Say About It,' in *Rapport fra lærebokkonferansen*, Caspar, Rådal.

Hoyles, C.: 1982, 'the Pupil's View of Mathematics Learning,' Educational Studies in Mathematics **13**.

Hua, L. K.: 1983, 'Some Experiences in Popularising Mathematical Methods,' in *Proceedings of the Fourth International Congress on Mathematics Education*, Birkhauser, Boston.

Hudson, K.: 1983, *The Language of the Teenage Revolution*, Macmillan, London.

Hymes, D.: 1980, 'On Communicative Competence,' in A. K. Pugh (ed.), *Language and Language Use*, Heinemann, London.

Høines, M.: 1983: *Begynneropplæringen i matematikk*, Caspar, Rådal.

IOWO: 1976, 'Five Years IOWO,' *Educational Studies in Mathematics* **7**.

Jahnke, N.: 1978, 'Zum Verhaltnis von Wissenentwicklung und Begrundung in der Mathematik — Beweisen als Didaktisches Problem,' *Materialen und Studien*, Universität Bielefeld.

Jenkins, R.: 1983, *Lads, Citizens and Ordinary Kids*, Routledge and Kegan Paul, London.

Jones, J.: 1974, 'Quantitative Concepts, Vernacular and Education in Papua New Guinea,' *Educational Research Unit Report*, 12, University of P.N.G.

Jones, P. L.: 1982, 'Learning Mathematics in a Second Language. A Problem with More and Less,' *Educational Studies in Mathematics* **13**.

Karabel, J. and Halsey, A. H. (eds.): 1977, *Power and Ideology in Education*, Oxford University Press, New York.

Keane, J.: 1983, 'Democracy and Ideology,' *Canadian Journal of Political and Social Theory* **7**.

Kemme, S. L.: 1981, 'References of Speech Acts as Characteristics of Mathematics Classroom Conversation,' *Educational Studies in Mathematics* **12**.

Kohl, H.: 1972, *36 Children*, Penguin, Harmondsworth.

Kozol, J.: 1968, *Death at an Early Age*, Penguin, Harmondsworth.

Krutetskii, V. A.: 1976, *The Psychology of Mathematical Abilities in School Children*, The University of Chicago Press, Chicago.

Kupert, A.: 1973, 'Introduction,' in G. Bateson, *Steps to an Ecology of Mind*, Paladin Books, Herts.

Labov, W.: 1972, *Language in the Inner City*, University of Pennsylvania Press, Philadelphia.

Labov, W.: 1972a, 'Some Principles of Linguistic Methodology,' *Language and Society* **1**.

Labov, W.: 1972b, 'On the Mechanism of Linguistic Change,' in J. J. Gumpertz and

D. Hymes (eds.), *Directions in Sociolinguistics*, Holt Rinehart and Winston, New York.
Laing, R. D.: 1968, *Det spaltete selv*, Gyldendal, Oslo.
Laing, R. D. and Cooper, D.: 1970, *Reason and Violence*, Tavistock Publ., London.
Lamb, S. M.: 1966, *Outline of Stratificational Grammar*, Georgetown University Press, Washington D.C.
Lancy, D. F. (ed.): 1978, 'The Indigenous Mathematics Project,' *Papua New Guinea Journal of Education*, Special Issue 14.
Land, F. W. and Bishop, A. J.: 1966—69, *Annual Reports of Mathematics Teaching Research Project*, The University of Hull, Hull.
Lean, G. and Clements, M. A.: 1981, 'Spatial Ability, Visual Imagery and Mathematical Performance,' *Educational Studies in Mathematics 12.*
Leonard, L. B.: 1976, *Meaning in Child Language: Issues in the Study of Early Semantic Development*, Grune & Stratton, New York.
Leont'ev, A. N.: 1978, *Activity, Consciousness and Personality*, Prentice Hall, Englewood Cliffs, New Jersey.
Leont'ev, A. N.: 1980, 'Tatigkeit und Kommunikation,' *Sowjetwissenschaft* **5**.
Leont'ev, A. N.: 1981, 'The Problem of Activity in Psychology,' in J. V. Wertsch (ed.), *The Concept of Activity in Soviet Psychology*, M. E. Sharpe, New York.
Lestage, A.: 1982, *Literacy and Illiteracy*, The Unesco Press, Paris.
Levi-Strauss, C.: 1985, *The Savage Mind*, The University of Chicago Press, Chicago.
Levine, K.: 1982, 'Functional Literacy: Fond Illusions and False Economics,' *Harvard Educational Review* **52**.
Liber, L. S.: 1981, 'Spatial Representation and Behaviour: Multiple Perspectives,' in L. S. Liber (ed.), *Spatial Representation and Behaviour across Life Span*, Academic Press, New York.
Lie, G.: 1981, *Gatelangs*, Universitetsforlaget, Oslo.
Lindquist, S.: 1979, *Grav dar du står*, Bonniers, Stockholm.
Lomow, B. F.: 1980, 'Die Kategorien Kommunikation und Tatigkeit in der Psychologie, *Sovietwissenschaft* **5**.
Løfsted, J. I.: 1980, *Chinese Educational Policy. Changes and Contradictions 1949—1979*, Almquist and Wicksell, Stockholm.
Mackie, R.: 1980, *Literacy and Revolution: The Pedagogy of Paulo Freire*, Pluto Press, London.
Maier, E.: 1980, Folk Mathematics, *Mathematics Teaching* **93**.
Malinowski, B.: 1923, 'The Problem of Meaning in Primitive Languages,' in C. Ogden and I. A. Richards, *The Meaning of Meaning*, Routledge and Kegan Paul, London.
Manger, P. (ed.): 1974, *Education in China*, Anglo-Chinese Education Inst., London.
Masuch, M.: 1974, *Uddannelsessektorens politiske økonomi*, Rhodos, Copenhagen.
Mead, G. H.: 1934, *Mind Self and Society*, Chicago University Press, Chicago.
Mead, G. H.: 1965, *On Social Psychology*, Chicago University Press, Chicago.
Mellin-Olsen, S.: 1977, *Læring som sosial prosess*, Gyldendal, Oslo.
Mellin-Olsen, S. (ed.): 1979, *Matematikken i elevenes omgivelser*, NKI, Oslo.
Mellin-Olsen, S.: 1980, *Didaktikk for integrering teori og arbeide*, Caspar, Rådal.
Mellin-Olsen, S.: 1981, 'Instrumentalism as an Educational Concept,' *Educational Studies In Mathematics* **12**.
Mellin-Olsen, S. (ed.): 1981a, *Rapport lærebokkonferansen 1981*, Caspar, Rådal.

Mellin-Olsen, S. and Rasmussen, R.: 1976, *Skolens vold*, Pax, Oslo.

Mills, C. W.: 1971, *The Sociological Imagination*, Penguin, Harmondsworth.

Mills, C. W.: 1974, 'The Cultural Apparatus,' in I. L. Horowitz (ed.), *Power, Politics & People. The Collected Essays of C. W. Mills*, Oxford University Press, London.

Mishler, E. G.: 1979, 'Meaning in Context: Is there Any Other Kind? *Harvard Educational Review* **51**.

Newell, A. and Simon, H. A.: 1972, *Human Problem Solving*, Prentice Hall, Englewood Cliff, New Jersey.

Newman, M. A.: 1977, 'An Analysis of Sixth-Grade Pupils' Errors on Written Arithmetical Tasks,' in M. A. Clements and J. Foyster (eds.), *Research in Education in Australia* **1**, University of Melbourne.

Niss, M.: 1983, 'Considerations and Experiences Concerning Integrated Courses in Mathematics and Other Subjects,' in *Proceedings of the Fourth International Congress on Mathematical Education*, Birkhauser, Boston.

McNeill, D.: 1970, *The Acquisition of Language*, Harper & Row, New York.

O'Neill, W. A.: 1970, 'Review on Schoolboys of Barbania. Letter to a Teacher,' *Harvard Educational Review* **40**.

Ogden, C. and Richards, I. A.: 1923, *The Meaning of Meaning*, Routledge and Kegan Paul, London.

Olson, D. R.: 1977, 'From Utterance to Text: The Bias of Language in Speech and Writing,' *Harvard Educational Review* **47**.

Otte, M.: 1984, 'Komplementaritet,' *Occasional Papers*, 42, IDM Universität Bielefeld.

Otte, M.: 1984a, 'Was ist Mathematik?' *Occasional Papers* 43, IDM, Universität Bielefeld.

Papert, S. 1980, *Mindstorms, Children, Computers and Powerful Ideas*, The Harvester Press, Brighton.

Piaget, J.: 1970, *Genetic Epistemology*, Columbia University Press, New York.

Piaget, J.: 1972, 'Intellectual Evolution from Adolescence to Adulthood,' *Human Development* **15**.

Pinxten, R.: 1983, *The Anthropology of Space*, The University of Pennsylvania Press, Philadelphia.

Plowden, L.: 1967, *Children and their Primary Schools*, Her Majesty's Stationery Office, London.

Prather, R. E.: 1982, *Problem-Solving Principles, Programming with Pascal*, Prentice Hall, Englewood Cliffs, New Jersey.

Reich, P. A.: 1970, *A Relational Network Model for Language Behavior*, University of Michigan Microfilms, Ann Arbor.

Reich, W.: 1969, *Character-Analysis*, The Noon-Day Press, New York.

Rosen, H.: 1972, *Language and Class: A Critical Look at the Theories of Basil Bernstein*, Falling Wall Press, London.

Ruesch, J. and Bateson, G.: 1951, *Communication: The Social Matrix of Psychiatry*, W.W. Norton, New York.

Sampson, G.: 1980, *School of Linguistics*, Hutchinson, London.

Saxe, G. B. and Posner, J.: 1983, 'The Development of Numerical Cross-Cultural Perspectives,' in H. Ginsburg (ed.), *The Development of Mathematical Thinking*, Academic Press, London.

Schaff, A.: 1973, *Language and Cognition*, McGraw-Hill, New York.

Sinclair, H.: 1983, 'Young Children's Acquisition of Language and Understanding Mathematics,' in Proceedings of The Fourth International Congress on Mathematical Education, Birkhauser, Boston.

Silberman, C. E.: 1970, *Crisis in the Classroom*, Random House, New York.

Skemp, R. R.: 1971, *The Psychology of Learning Mathematics*, Penguin, Harmondsworth.

Skemp, R. R.: 1976, *Intelligence, Learning and Action*, J. Wiley, London.

Smedslund, J.: 1977, 'Piaget,' *Norsk Pedagogisk Tidsskrift* **61**.

Smitherman, G.: 1981, ' "What Go Round Come Round": King in Perspective.' *Harvard Educational Review* **51**.

Stone, M.: 1981, *The Education of the Black Child in Britain*, Fontana Books, Glasgow.

Stubbs, M.: 1976, *Language, Schools and Classroom*, Methuen, London.

Stubbs, M.: 1976a, *Explorations in Classroom Observation*, J. Wiley, London.

Stubbs, M.: 1980, *Language and Literacy, The Sociolinguistics of Reading and Writing*, Routledge and Kegan Paul, London.

Swetz, F. and Ying-King Yu: 1979, 'Mathematical Olympiads in the People's Republic of China,' *Educational Studies in Mathematics* **10**.

Swetz, F.: 1978, *Socialist Mathematics Education*, Burgundy Press, Southampton, USA.

Teitelbaum, H. and Hiller, J.: 1977, 'Bilingual Education: The Legal Mandata,' *Harvard Educational Review* **49**.

Thompson, J. B.: 1983, 'Ideology and Domination' *Canadian Journal of Political and Social Theory* **7**.

UNESCO: 1981, *Youth in the 1980s*, the Unesco Press, Paris.

Volosinov, V. M.: 1976, *Freudianism: A Marxist Critique*, Academic Press, New York.

Vygotsky, L.: 1962, *Thought and Language*, The MIT Press, Cambridge, Mass.

Vygotsky, L.: 1979, *Mind in Society*, Harvard University Press, Cambridge, Mass.

Wertsch J. V.: 1981, 'Introduction,' in J. Wertsch (ed.), *The Problem of Activity in Soviet Psychology*, M.E. Sharpe, New York.

Whorf, B. L.: 1967, *Language, Thought and Reality*, The MIT Press, Cambridge, Mass.

Wilden, A.: 1968, *The Language of the Self*, Johns Hopkins Press, Baltimore.

Wilden, A.: 1972, *System and Structure*, Tavistock Publ., London.

Willis, P.: 1976, 'The Class Significance of School Counter-Culture,' in M. Hammersley (ed.), *The Process of Schooling*, Routledge and Kegan Paul, London.

Willis, P.: 1977, *Learning to Labour*, Saxon House, Farnborough.

Willis, P.: 1981, 'Cultural Production is Different from Cultural Reproduction is Different from Social Reproduction is Different from Social Reproduction is Different from Reproduction,' *Interchange* **12**.

Woods, P.: 1983, *Sociology and the School*, Routledge and Kegan Paul, London.

INDEX OF NAMES

INDEX OF SUBJECTS

MATHEMATICS EDUCATION LIBRARY

Managing Editor:
A. J. BISHOP, *Cambridge, U.K.*

Hans Freudenthal, *Didactical Phenomenology of Mathematical Structures*. x + 595 pp., 1983.

B. Christiansen, A. G. Howson, and M. Otte (eds.), *Perspectives on Mathematics Education*. xii + 371 pp., 1986.

Adrian Treffers, *Three Dimensions. A Model of Goal and Theory Description in Mathematics Instruction – The Wiskobas Project*. xvi + 351 pp., 1987.

Stieg Mellin-Olsen, *The Politics of Mathematics Education*. xvi + 249 pp., 1987.